T5-AGA-233

ENVIRONMENTAL DECISION SUPPORT SYSTEMS

ELLIS HORWOOD SERIES IN COMPUTERS AND THEIR APPLICATIONS

Series Editor: BRIAN L. MEEK, Director of Information Technology, Goldsmiths' College, London, on secondment from King's College London

Abramsky, S. & Hankin, C.J. ABSTRACT INTERPRETATION OF DECLARATIVE LANGUAGES
Alexander, H. FORMALLY-BASED TOOLS & TECHNIQUES FOR HUMAN–COMPUTER DIALOGUES
Atherton, R. STRUCTURED PROGRAMMING WITH BBC BASIC
Atherton, R. STRUCTURED PROGRAMMING WITH COMAL
Bailey, R. FUNCTIONAL PROGRAMMING WITH HOPE
Barrett, R., Ramsay, A. & Sloman, A. POP-11
Bharath, R. COMPUTERS AND GRAPH THEORY
Bishop, P. FIFTH GENERATION COMPUTERS
Bullinger, H.-J. & Gunzenhauser, H. SOFTWARE ERGONOMICS
Burns, A. NEW INFORMATION TECHNOLOGY
Carberry, J.C. COBOL
Chivers, I.D. AN INTRODUCTION TO STANDARD PASCAL
Chivers, I.D. MODULA 2
Chivers, I.D. & Clark, M.W. INTERACTIVE FORTRAN 77
Clark, M.W. PC-PORTABLE FORTRAN
Clark, M.W. TEX
Cope, T. COMPUTING USING BASIC
Curth, M.A. & Edelmann, H. APL
Dahlstrand, I SOFTWARE PORTABILITY AND STANDARDS
Eastlake, J.J. A STRUCTURED APPROACH TO COMPUTER STRATEGY
Eisenbach, S. FUNCTIONAL PROGRAMMING
Ellis, D. MEDICAL COMPUTING AND APPLICATIONS
Ennals, J.R. ARTIFICIAL INTELLIGENCE
Ennals, J.R BEGINNING MICRO-PROLOG
Ennals, J.R., *et al.* INFORMATION TECHNOLOGY AND EDUCATION
Fernando, C. CLUSTER ANALYSIS ON MICROCOMPUTERS
Filipič, B. PROLOG USER'S HANDBOOK
Guariso, G. & Werthner, H. ENVIRONMENTAL DECISION SUPPORT SYSTEMS
Harland, D.M. CONCURRENCY AND PROGRAMMING LANGUAGES
Harland, D.M. POLYMORPHIC PROGRAMMING LANGUAGES
Harland, D.M. REKURSIV
Harris, D.J. DEVELOPING DEDICATED DBASE SYSTEMS
Hepburn, P.H. FURTHER PROGRAMMING IN PROLOG
Hepburn, P.H. PROGRAMMING MICRO-PROLOG MADE SIMPLE
Hill, I.D. & Meek, B.L. PROGRAMMING LANGUAGE STANDARISATION
Hutchins, W.J. MACHINE TRANSLATION
Hutchison, D. FUNDAMENTALS OF COMPUTER LOGIC
Hutchison, D. & Silvester, P. COMPUTER LOGIC
Koopman, P. STACK ORIENTED COMPUTERS
Koster, C.H.A. TOP-DOWN PROGRAMMING WITH ELAN
Last, R. AI TECHNIQUES IN COMPUTER ASSISTED LANGUAGE LEARNING
Lester, C. DATA STRUCTURES
Lucas, R. DATABASE APPLICATIONS USING PROLOG
Lucas, A. DESKTOP PUBLISHING
Matthews, J.L. FORTH
Millington, D. SYSTEMS ANALYSIS AND DESIGN FOR COMPUTER APPLICATIONS
Moore, L. MASTERING THE ZX SPECTRUM
Moseley, L.G., Sharp, J.A. & Salenieks, P. PASCAL IN PRACTICE
Moylan, P. ASSEMBLY LANGUAGE FOR ENGINEERS
Narayanan, A. & Sharkey, N.E. AN INTRODUCTION TO LISP
Parrington, N. & Roper, M. UNDERSTANDING SOFTWARE TESTING
Paterson, A. OFFICE SYSTEMS
Phillips, C. & Cornelius, B.J. COMPUTATIONAL NUMERICAL METHODS
Rahtz, S.P.Q. INFORMATION TECHNOLOGY IN THE HUMANITIES
Ramsden, E. MICROCOMPUTERS IN EDUCATION 2

Series continued at back of book

ENVIRONMENTAL DECISION SUPPORT SYSTEMS

G. GUARISO
Department of Electronics — Research Centre for Environmental Computer Science (CIRITA), Politecnico di Milano, Italy
H. WERTHNER
Institute of Statistics and Computer Science
University of Vienna

ELLIS HORWOOD LIMITED
Publishers · Chichester

Halsted Press: a division of
JOHN WILEY & SONS
New York · Chichester · Brisbane · Toronto

First published in 1989 by
ELLIS HORWOOD LIMITED
Market Cross House, Cooper Street,
Chichester, West Sussex, PO19 1EB, England
The publisher's colophon is reproduced from James Gillison's drawing of the ancient Market Cross, Chichester.

Distributors:
Australia and New Zealand:
JACARANDA WILEY LIMITED
GPO Box 859, Brisbane, Queensland 4001, Australia
Canada:
JOHN WILEY & SONS CANADA LIMITED
22 Worcester Road, Rexdale, Ontario, Canada
Europe and Africa:
JOHN WILEY & SONS LIMITED
Baffins Lane, Chichester, West Sussex, England
North and South America and the rest of the world:
Halsted Press: a division of
JOHN WILEY & SONS
605 Third Avenue, New York, NY 10158, USA
South-East Asia
JOHN WILEY & SONS (SEA) PTE LIMITED
37 Jalan Pemimpin # 05–04
Block B, Union Industrial Building, Singapore 2057
Indian Subcontinent
WILEY EASTERN LIMITED
4835/24 Ansari Road
Daryaganj, New Delhi 110002, India

British Library Cataloguing in Publication Data
Guariso, Gorgio
Environmental decision support systems. —
(Ellis Horwood books in computing science.
Series in computers and their applications)
1. Environmental planning. Applications in computer systems
I. Title II. Werthner, H.
711'.028

Library of Congress Card No. 89–1060

ISBN 0–7458–0255–9 (Ellis Horwood Limited)
ISBN 0–470–21431–7 (Halsted Press)

Typeset in Times by Ellis Horwood Limited

Printed and bound in Great Britain at
The Camelot Press Ltd, Southampton

Table of Contents

To Stella, the star of the South.

To Claudia and Gabriele, and all children,
may they still have a lively environment.

Preface

The discussion about decision support systems (DSS) can be considered, in terms of time spans in computer science, a long-lasting one. It has been going on for ten years, providing several different definitions of such systems. This discussion will certainly continue, enriched by the experience gained through applications of this concept and by new insights from other fields. Even while writing this book, the authors were confronted with several new experiences and opinions, and the basic structure of the book underwent quite an evolution as a result.

The main purpose of environmental decision support systems (EDSS) is to support decision makers in solving problems that are poorly or insufficiently structured, as is often the case in environmental situations. An EDSS should provide a combination of several tools necessary to support the process of structuring a problem, to gain new insights about it, to look for examples of problems that have already been solved, to produce alternative solutions and to evaluate them. The scope of a DSS is not to replace specific human capacities but to support users. In the end, the time and the steps necessary to find a satisfying solution to a problem should be essentially shortened. It follows from these very important objectives that the approaches to satisfying them can be very different and often controversial. Thus, the authors have tried to narrow the range of possible alternatives and to apply them to environmental problems.

In the field of environmental management, information technology has passed from rather restricted calculation applications to the wide area of information processing. At the beginning, mathematical modelling in the sense of 'classical' number crunching was used nearly exclusively, whereas nowadays database and also artificial intelligence techniques are applied. The authors were confronted with one basic problem: the concept of EDSS described in the following chapters cannot be found in a pure form in actual applications in the field of environmental management. Thus another approach has been chosen to explain the concept. The necessary basic tools applied in environmental programs are described and analysed in several examples ('learning by example'); our method has been to isolate from several computer applications the features necessary to build an EDSS. Finally, the authors present a prototype in which these features and techniques are integrated. It is to be hoped that this method will prove to be a fruitful one, giving some new impulse to future work.

The use of computers is not, strictly speaking, a prerequisite for solving problems in the field of environmental management and planning. However, the computer is becoming more and more an essential tool with which to tackle and to solve, in a more satisfying way, problems which have been left until recently to the experience and common sense of political and technical decision makers. Thus, this book will deal more with the design of computer systems for this kind of application than with decision or behavioural theories, on which the general concepts of DSS and the models they use are usually based. This book is not exclusively technical, however, and the authors have embedded into it a rather general framework with a description of the historical roots of DSS and computer applications in environmental management. Since DSS is seen here as an integration of some basic modules, such as a database or a modelbase, parts of the book serve as an introduction to certain branches of computer science. Latest software and hardware achievements are not discussed in detail because of the fast progress in that area, while some basic concepts are developed which are not expected to change rapidly and which may help in understanding and constructing EDSS. Thus, while the book is on a rather conceptual level, several implementational points are briefly described where necessary.

The interdisciplinary character of an EDSS, in both construction and use, should be stressed. People involved in these tasks may range from environmental experts (chemical and sanitary engineers, ecologists, etc.) to systems analysts (dealing with modelling, systems theory and operations research) and computer experts (for data management, software design or choice of hardware). Thus this book is directed towards people working in environmental agencies, consulting firms or research centres who want to start using (or improve the way they are using) a computer. Other potential readers might be computer experts seeking a review of the main features and conditions of effective computer projects in this area.

A few comments on the context of computer applications in organizations, whether they are dealing with environmental problems or not, seem to be necessary. Information system projects nearly always begin with basic assumptions which are not explicitly spelled out, and this is often the reason for their failure or for long delays. It is thus important to understand the type of rationality which exists inside an organization and which may be supported by computer systems.

Several approaches to viewing organizational rationality can be distinguished and should be identified before constructing a distinct application (see also Ciborra 1988):

— *Unlimited rationality and mechanistic knowledge*: In this case, organizational knowledge is fully accessible and follows rationality axioms. This can be seen as the oldest existing means of explaining the functioning of an organization. Modelled systems are seen as a direct mapping of reality, decisions can be based on hard facts and follow rather deterministic rules. In this context a DSS is mainly based on normative models.

— *Limited rationality and tacit knowledge:* Designers can be described as persons with limited capabilities (Simon 1976), which also limits the design of a system. It is assumed that knowledge cannot be completely formalized, that models only respect a limited set of variables and that designs are incomplete. The scope is not optimality but sufficiency. Two drawbacks of this organizational perspective can be mentioned: it ignores conflicts of interest and related opportunistic behaviour, and it relies on cognitive processes that ignore cognitive biases.
— *Strategic rationality and opportunistic information:* As soon as organizations are viewed as mixed interest groups, opportunistic knowledge plays a role. Knowledge can be manipulated. A transaction cost approach, borrowed from the field of economics, can provide an explanation and some understanding of information processing in organizations. It reflects bounded rationality and the opportunism of a decision maker. An organization can be seen as a mixture of conflict and cooperation. Therefore, the systems that are designed have to support negotiations. The concept of a DSS changes from planning and control to communications and contracting.
— *Adaptive rationality and limited learning:* This concept involves the problems of organizational changes. A set of pre-existing arrangements or frames can also be viewed as the formative context which determines the functioning of an institution. An information system has to support changes in activities and not only execute existing tasks. Repeated loops of learning are necessary and have to be supported. Such loops should facilitate incremental changes in routines and organizational values.

Different forms of organizations, either strictly hierarchical or rather cooperative, are correlated to these isolated types of rationality. A DSS builder should try to analyse the functioning of an institution and to respect such conditions and circumstances.

Furthermore, there is a danger of restrictiveness when somebody constructs or uses an EDSS (as any other system), as it may exclude some types of decision making procedures and it may prejudge, by the choice of certain parameters, possible outcomes. Nevertheless, a DSS may increase the effectiveness of decision making and may also lead to new decisions because more objective data and procedures are made available.

It follows that an EDSS has to be seen as an open system, which should contain the possibility of learning on the part of both the system and the human user. A DSS architecture should enable developments in several directions rather than be an obstacle to changing the structure of an organization. And, by making the data and the procedure on which a decision is based explicit, it may also lead to more objectivity.

The structure of the book is as follows. Chapter 1 consists of historical background and an overview of the development from traditional software applications to DSS. Different approaches to defining the concept of a DSS are considered and discussed. Objectives of such a system are isolated and embedded into a more general framework. Finally, three software architectures are presented, and the different basic modules are described.

Chapter 2 deals with the basic criteria for applying computer systems to environmental problems. It also contains a short history of computers and mathematical modelling in the field. A more phenomenological or practical approach is used to take into account specific features of environmental systems and problems, such as the complexity of the physical as well as the decision structure, large geographical areas, large amounts of data or dynamic behaviour. This chapter specifies the more general characteristics of DSS that were developed in the first chapter for environmental decision making.

The following chapters deal with the specific components of a DSS, following the architecture developed in Chapters 1 and 2. All of them begin with an introduction to the specific field of computer science in order to make the text readable for non-computer experts. Chapter 3 discusses the database component of a DSS. It contains such topics as different data models, their application to environmental problems, geographical information systems and the possibility of their integration with other EDSS modules.

Chapter 4 concerns the role played by mathematical models. Different types of models and their usefulness are discussed: models for environmental planning, control, forecast, and simulation. The specific importance of simulation models in environmental decision making is underlined.

Chapter 5 covers the artificial intelligence part of decision support systems and points out the various ways in which it may be interrelated with the mathematical models and the database component. Different techniques, such as first-order logic and frames, are presented and integrated. These methods may serve both for knowledge representation and for control of an EDSS by an inference mechanism.

The design of a user interface and the possible techniques for its construction are the object of Chapter 6. This is closely related to certain hardware aspects, as they constitute the 'hard' conditions for an interface design. This chapter is rather technical in character but is by no means a complete description of the various possibilities, since the development of hardware devices is extremely fast.

In the final chapter, a prototype is presented which incorporates all of the necessary features in the proposed architecture of an EDSS. This system is the product of work in a software laboratory and is not yet in practical use.

All but the first chapter give several practical examples and case studies illustrating the specific problems dealt with.

REFERENCES

Ciborra, C. U. (1988) Knowledge and systems in organizations. In: Lee, R. M., McCosh, A. M., & Migliarese, P. (eds), *Organizational Decision*

Support Systems. Proc. of the IFIP WG 8.3 Working Conference, Como, 20–22 June, 1988, pp. 229–246.

Simon, H. (1976) *Administrative Behavior*. The Free Press, New York.

ACKNOWLEDGEMENTS

The authors would like to thank their home institutions, i.e. the Department of Electronics of the Politecnico di Milano and the Institute of Statistics and Computer Science of the University of Vienna, as well as the Austrian Schrödinger foundation and the 'Centro Teoria dei Sistemi — CNR'.

The emergence of this book was influenced by our contacts with many colleagues. The examples in the respective chapters are the result of our cooperation with them; they also contributed to the central concepts of the book by numerous discussions. Those who contributed include I. Bomze, A. Colorni, G. Finzi, K. Fröschl, C. Gandolfi, M. Gatto, W. Grossmann, M. Hitz, Z. Jurkiewicz, G. Kappel, A. Kraszewski, E. Laniedo, S. Rinaldi, R. Soncini-Sessa and A. Tjoa. Our students in the academic year 1987–88, F. Melodi, M. Moscioni, L. Nisoli and A. Rizzoli, did a lot of programming in the LISP and LOOPS environment and developed parts of the software described in the book. Our special thanks go to C. Neuenschwander-Unterleitner; she was extremely helpful in reading the manuscript.

H. W.
Vienna, Austria

G. G.
Milan, Italy

1

A decision support systems framework

The purpose of this chapter is to present the reader with an introduction to decision support systems (DSS) based on a discussion of technical arguments as well as on historical background. The development of DSS, and its origins in computer science, will be traced and a general framework, and some proposed architectures, will be presented. It should be pointed out, however, that a detailed description of these complex circumstances lies beyond the scope of the present work.

Modern sciences, especially those related to computing machines, have to be seen as products of several trends in society, which are often conflicting. The development of new fields is not a straight and linear process; different disciplines may break away or combine to form new sciences. Decision support systems are a product of this development and represent an approach which tries to integrate in an interdisciplinary manner different fields of computer science as well as other disciplines. Moreover, DSS might evolve from a wish to manage the complex situations that exist in modern society. Thus, while the image of DSS is not very clear and also rather heterogeneous, this feature makes them attractive for both users and researchers.

The present chapter is a mixture of sociological reflection and technical introduction. It includes a review of developments in society and in the field of decision making and the progress in computer science. Although the authors are aware of the danger of remaining on a phenomenological level, they have attempted to describe these different roots to provide a deeper understanding.

In Chapter 1, different ways of defining and approaching DSS are discussed; specific features for their application in environmental management will be developed in Chapter 2. Section 1.1 is a review of technological progress; section 1.2 describes different approaches to the term DSS. Sections 1.3 and 1.4 present the evolution which led to DSS and give a tentative definition of such systems. In section 1.5, several architectures of DSS, which have been proposed in the literature, will be described and discussed. After an evaluation of these concepts, a configuration will be proposed which fulfils the diverse needs that have been identified. The final section also refers to the context in which DSS can be seen.

1.1 TECHNOLOGICAL PROGRESS

Today mankind is witnessing rapidly evolving technological progress; a major feature of this progress is the steady decrease in the time between the introduction of new inventions and products and their replacement by even newer products. A brief look at the field of microcomputers may illustrate this process. Eight-bit processors, which formed the basis for modern personal computers at the end of the seventies, which radically changed the habits of computer scientists and led also to a broad diffusion of these machines, are now outmoded and have already been replaced by 32-bit personal computers. Such frontiers as limited memory capacity and speed of 'early' personal computing have been passed over within a period of ten years.

Confronted with a rapidly growing number of scientific insights, technological changes and new products, the potential user has the problem of monitoring developments and using them for his/her own needs. Even the specialist has problems in remaining abreast of innovations in the field and evaluating seriously all or only a major part of them. A similar problem arises in fields of science where a great deal of specialization has taken and still is taking place. Bibliographical databases are nearly exploding, and it is quite impossible to read all new publications in even one area of research. An original description of the problem of keeping up with publications in a limited field is given in Schwendter (1982).

The invention of the computing machine marks a significant moment in history. It can be seen as the intersection of two technological developments, and in each of these the computer has had revolutionary effects. In the first area, which has to do with human labour and its material conditions, the computer and its basic technology represent the so-called third industrial–technological revolution.†

The second development concerns the role of human beings as symbol and information processors. The computer, which is, in contrast to all former machines, a symbol manipulator and transformer, partially models and simulates of human mental capacities; thus, its invention can also be called a revolution in information processing. So, the computer stands at the intersection of two developments, which drastically changed the physical and mental relationship of a person to his/her environment.

This technological and social progress, of which the computer forms a part, was not straightforward. It was accompanied by great disturbances and contradictions. Together with other factors, it produced overall material prosperity and, simultaneously, regional and individual poverty. This development also resulted in an ecological situation not only potentially dangerous to the whole of mankind but also characterized by growing social

† The first revolution, which also marked the beginning of modern society, took place on three different levels: (a) mechanical systems replaced human capabilities; (b) inanimate force, especially the steam engine, replaced human and animal force; and (c) production processes were radically improved (see Landes 1968). The second revolution was set in motion by the invention of the motor car and electrical power.

fractures and sometimes accompanied by a paralysis in decision making.† The changes in the political situation in some countries in the western world over the last decade are also influenced by this circumstance.

Organization and complexity

Modern society can best be described with the help of the term 'complexity'. In trying to identify single aspects of society by using a social, economic, ecological, or cultural point of view, one realizes the permanent tendency towards separation with a simultaneous growth of interdependencies. Phenomenologically, large-scale and international organizations can be taken as an example of this process. They might be subdivided into national, regional or other divisions, each existing and working independently, but on an overall level they have manifold and strong connections.

In general, there seems to be a correlation between the growth of social entities (i.e. organizations, firms, institutions or universities), on the one hand, and complexity, on the other. From a historical point of view, large-scale organizations with a parallel high degree of interrelationships represent a new development. And, as in the case of the above-mentioned progress in technology and information processing, they are at the same time products and causes of industrial changes. Another important feature should be mentioned in connection with this development. Large organizations are also large information processing systems. The digestion of information is one of the preconditions for their functioning. In fact, the work of some of them is predominantly in information processing. Examples of this latter type are news organizations, computer firms, banks and the United Nations organizations.

Because of the importance of information processing, some scientists, for example Bell (1973), call the present historical period already a 'postindustrial society' or 'the information age'. With this central role in mind,‡ they go so far as to call information a main resource, as important as (or even exceeding) human labour and capital.§ In any case, it is clear that the complexity of today's society is correlated with information processing machinery.

Man–machine systems

Modern society can also be characterized by another property, namely the importance or even dominance of man–machine systems. Nearly all interactions between a person and his or her environment are carried out via or with the help of instruments. It seems adequate to use the term 'system' for this

† It is not within the scope of the book to discuss the complex relation of social and technological development, but engineers should be aware of the fact that they sometimes might be the subject, and not merely the object, of progress.

‡ As a recent example, consider the stock market crash of 1987 which was produced, among other means, by errors in information processing (IEEE Spectrum 1987).

§ Statistically, their thesis is supported by the transfer of human labour from the agricultural and industrial economic sector to the service sector and the rapid growth of white-collar or knowledge workers.

type of interaction. Historically, the mechanical parts of a system had a subordinate role; they were controlled by the human. But technology liberated human beings from certain physical tasks and efforts. In actual man–machine systems, for both material and information processing, machines play a more important role and are no longer restricted to physical work. They may also be responsible for controlling tasks in the complicated interaction between humans and the environment. In fact, the impression might be given that machines might, sooner or later, liberate themselves from control by human beings.

This latter development is caused prominently by the computer. A computer can store, display and transform information, therefore differing significantly from all other information-handling machines (i.e. telephones, television, photography, etc.). Its ability to transform data from one representation to another, to store the description of a data transformation in the same way and form as the data themselves, gives the computer the power to model partially the human brain and some of its tasks. Originally developed for freeing human beings from repetitive calculation tasks (Goldstine 1971), its capacity for storing vast amounts of data and for controlling the work of other machines was soon recognized. In a sense a computer can be seen as a general-purpose machine which needs only some extremities for fulfilling specific tasks.

Obviously, computers are more than very fast calculators for repetitive or complex operations and storage devices for vast amounts of data. With progress in such fields as Artificial Intelligence (AI), they show even more significantly their capacity to support or to replace humans in their traditional area of creative or analytical thinking. Even if the power of AI products is still limited and far less than that promoted by some vendors, such products can give some indication about the future progress of man–machine systems. Products such as Expert Systems and decision support systems are in that sense a significant keystone of the evolution of the complexity of today's society and technological progress. Thus, we are confronted with a rather strange situation: such systems are the product of a complex history, from both a social and a technical point of view. At the same time, they seem to be one of the possible tools to manage one result of this development, namely the complexity of modern society. DSS are not these general problem solvers but, as a part of a sophisticated man–machine system, the importance here being placed on 'man' rather than on 'machine', they may constitute a small step in the right direction. This is the background against which the use of computer-based decision support systems in the field of environmental management will be presented and described.

1.2 APPROACHES TO DSS

Because of the general nature of the term 'decision support system' and the wide range of possible interpretations and definitions, a different approach will be taken to characterize and describe DSS. Thus, the following discussion centres on aspects, properties and concepts which are related to decision making and decision support and their automation.

DSS is often described as an interactive computer-based system that helps decision makers to utilize data and models in the solution of unstructured problems. This description, which refers to interactivity, lack of structure and a combination of data and models, proves to be rather restrictive in that all of the features mentioned are to be found in only a small number of existing systems. On the other hand, as in general all computer applications help in the decision making process in some way, nearly all of them could be called DSS. Both definitions, the restrictive and the broad, are too general to satisfy the need to understand, evaluate and design such systems. The problem of defining DSS, the possibility of seeing them as being all or nothing, accompanies the discussion about DSS (Keen 1987).

A list of necessary characteristics of a DSS was drawn up by Parker and Al-Utabi (1986), after reviewing 350 papers related to the subject. According to them, a DSS should

— assist managers in their decision making process for unstructured or semi-structured tasks,
— support and enhance rather than replace managerial judgement,
— improve the effectiveness of decision making rather than the efficiency, putting more weight on a correct decision than on the time it takes to make that decision,
— combine the use of models or analytical techniques with data access functions,
— emphasize flexibility and adaptability to respect changes in the context of the decision process, and
— focus on features which make them easy to be used interactively by non-experienced users.

Supplementary DSS can be distinguished from related fields such as Operations Research or Management Science, which rely more explicitly on normative models, through the following properties:

— enabling an intuitive approach towards a solution,
— helping in tentative procedures, as they could be supported by fast, prototyping environments,
— including trial-and-error procedures, and
— allowing the introduction of subjective judgements.

What does the term 'structure' mean?

DSS are employed nearly exclusively, as mentioned in all of the current literature, with semi-structured or unstructured problems or problem domains. It is thus necessary to differentiate structured from unstructured problems, together with the correlated necessary human interventions.

If the decision process can be well formulated in an algorithmic way, a computer program can, to some extent, take the part of the decision maker. Decisions in this case are unambiguous because alternatives and solution procedures are known. On the other hand, when the formulation becomes more vague, or the decision has to be made in a new context, or the problem is too complex for existing algorithms, the human processing capability becomes more important. An unstructured problem may be caused, for example, by lack of data or knowledge or by variables that are not quantified or by too great complexity. In this case, human intervention is necessary. For example, it is impossible to describe the movements of a person at a molecular level because of their complexity, but one can approximate them with a simpler physical model of motion at a higher level. In this case, it seems to be obvious that the structuring of the problem, i.e. the choice of the level of description, must be performed by a human being. Being a deterministic machine, the computer at the final end is an instrument for solving well-structured and defined problems. An unstructured problem, with respect to the above-mentioned causes, has to be resolved mainly by the human in the man–machine system. But there may be implemented algorithms in the system to support this process, resulting in a sequence of well-structured tasks.

Structured Problem	⟵⟶	Unstructured Problem
Strategies can be externalized as computer programs	Strategies have yet to be externalized but are readily applied to common problems	General problem-solving strategies must be used to deal with uncommon problems: — use of analogy — problem redefinition — intuition — formulation of specific strategies from existing ones — approximation

Fig. 1.1 — The Structure Continuum in the decision process (see Bonczek *et al.* 1981).

Fig. 1.1 shows the spectrum of the decision process. As the boundary moves to the right, from structured to unstructured, more general solution strategies in the man–machine procedure have to be used. These may include: (a) the use of analogy to find similar, already well-known problems; (b) the redefinition of the problem with different, but known, terms; (c) the deduction of a particular strategy from an existing one; (d) the use of intuitive approaches; and (e) approximating a problem using another at a level that is easier to describe.

Another view with respect to this differentiation is presented by Lewandowski and Wierzbicki (1987), who also look at the decision process. Areas in which human creativity may be replaced by automatic procedures, such as the automatic control of different industrial processes, are identified. These involve repetitive decisions which have to be made very fast and precisely and can be made rather autonomously by the machine. On the other hand, there are many decision making situations in which human insights and judgements are indispensable. The role of a DSS is to support a human in this area, the human still being the more important part of the man–machine system. In this case, the computerized system may provide tools for structuring a problem, such as by responding to 'if . . . then' questions, providing fast access to data or modelling capacities.

What does the term 'problem' mean?

Up to this point the descriptions or approaches to a DSS have been based on situations in which there is a clear problem definition or a clear awareness of problematic situations by all actors in a decision process. DSS serve to solve such problems. Some authors, such as Landry *et al.* (1985), discuss the concept of a 'problem' in the context of DSS in depth and confront two theoretical perspectives:

— problems viewed as 'objective realities', or
— problems seen as 'mental constructs'.

In the first case, problems are viewed in a pragmatic way as unsatisfactory objective realities discovered by observations and facts. Within this perspective, the analyst focuses on the facts as objective existing realities. The designer or expert has to observe, define and formulate the problem and its domain. As a problem exists objectively, all participants see it in the same way, even if there are different solutions. In this sense, the problem definition is a rather preliminary step to DSS design. This approach analyses the procedure to be taken when a problem and its domain have been well analysed and defined. Additional characteristics are connected with this approach: mainly normative models are used; and the complete set of possible alternatives is well known.

An alternative view considers a problem to be rather a subjective presentation conceived by an actor confronted with a reality which he or she sees and perceives to be unsatisfactory. The subjective dimension is recognized and a problem becomes 'someone's' problem. This point of view can also be closely correlated with the observations made at the beginning of this chapter about a complex socio-economic situation which also leads to a fracture in the perception of this situation. Consider, as an example, the different perception of pollution by individuals as children and as adults. Here, first of all, common threshold values have to be defined by the different actors before some other procedure can take place.

This relativistic approach leads to a stronger integration of problem formulation techniques into the context of a DSS and also stresses the major role of humans in the complex man–machine system. The emphasis is on the design and not on the analysis phase. Moreover, it sees decision finding more as a social process rather than as a technical, resolvable problem.

This social dimension of decision finding will not be discussed in the following. However, a common understanding is found during some social interaction between the decision making actors or institutions. The technical system does not have the scope to replace these complex processes, although it may be able to support them.

1.3 EVOLUTION OF DSS

DSS can be embedded in the development of such different fields as computer science and management science. One can distinguish between an evolutionary and a theoretical characterization of DSS with respect to Management Information Systems (MIS), as is done for example by Sprague and Carlson (1982).

In the evolutionary characterization, development is traced from Electronic Data Processing (EDP) at the beginning of the modern information processing age to DSS, which are handled as a logical advancement in this process. By distinguishing between features of EDP, MIS and DSS, a rather linear, straightforward progression may be drawn.

The basic characteristics of EDP are:

— focus on data, storage and processing at the operational level,
— efficient transaction processing,
— summary reports for management.

MIS focuses on information activities and intelligent data-retrieval functions. Characteristics include:

— information focus, mainly for middle management,
— integration of EDP jobs by business functions,
— retrieval and report function.

With respect to this characterization DSS have a higher position inside an institution and exhibit the following features:

— decision-focused, mainly for strategic and long-term decisions,
— flexibility and user friendliness,
— support of personal decision making styles.

Under this evolutionary aspect, the different levels inside an organization where these systems are located can be identified, but this might be misleading. It implies that decision making is located at the top level, thus assuming implicitly a strict hierarchical structure inside an organization and ignoring the different levels and ranges of the decision making process. And some may not agree with that narrow definition or description of MIS.

In the theoretical characterization of MIS, the overall context of information systems inside an organization is respected. The overall objective of MIS as quoted by Sprague and Carlson (1982) is 'to improve the performance of knowledge workers in institutions through the application of information technology'.

According to Head (1967) the pyramid of Fig. 1.2 may be used to describe MIS and DSS in a broader sense. The vertical dimension represents the levels of management and the horizontal dimension the main functional

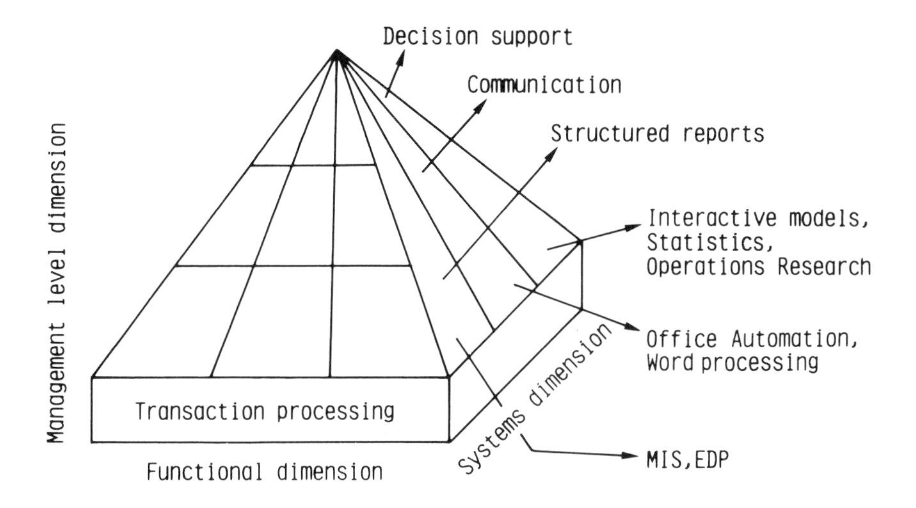

Fig. 1.2 — Embedding of decision support (see also Sprague and Carlson 1982).

areas of an institution. Transaction processing may be added as a base of the system. In the depth dimension some technologies are illustrated, which support the work of managers and other knowledge workers.

The three divisions in the depth dimension represent only one possible set of technologies: others could be added. The reporting system contains reports needed for the control and management of the institution as well as for general information needs. It has evolved from efforts in EDP and MIS. Communication systems result from progress in telecommunication and computer networks. All kinds of office information systems (from simple word processors to integrated office-support tools) show a strong influence. Based on these observations DSS seem to evolve from efforts in the field of Operations Research, Management Science and decision making.

From the theoretical point of view, DSS constitute more than an evolution beyond MIS. The main focus of MIS is on information, whereas DSS are concentrated rather on decision support. MIS and other EDP functions will not lose their importance in the organizational context. But, especially with respect to the decision finding and modelling aspect, DSS exhibit new features. They are not merely data retrieval or aggregation systems; their modelling, evaluation, and judgement abilities make the difference.

1.4 DEVELOPING OBJECTIVES OF DSS

Levels of decision making

In the previous section, a point of view was presented, which places a DSS and its use at the top of a hierarchy inside an institution. Underlying a hierarchic organizational structure, Anthony's (1965) taxonomy draws a distinction between several functional areas of decision making and thus places it at different levels, namely:

— strategic planning,
— management control, and
— operational control.

Strategic planning has to do with decisions about the objectives or changes in the objectives of an institution. Overall, long-term policies are set at this level. *Management control* is a combination of planning and control. Its main scope is the utilization and allocation of resources in an effective way. At this level, human judgement is still of significant importance. Finally, *operational control* has to guarantee efficiency in performing operations. It is controlled by specific rules, chosen at a higher level. Operational control has short time ranges.

In general, it seems to be difficult to differentiate between planning and controlling functions. Rather, they should be represented by a continuum, with strategic planning at one end, management control in the centre and operational control at the other end. Anthony's taxonomy fits in well with the distinction between structured and unstructured problems as described above. Operational control decisions seem to be highly structured and appear frequently. Thus, algorithmic or fixed sets of solution rules exist and are used. At the other end of the continuum, strategic planning decisions are taken less frequently; often the problem domain is not explicitly defined and data are less likely to be available. Specific solution strategies and decision procedures may not be defined. This does not mean that there are no formal methods for making decisions at the strategic level, but human capabilities, for example associative reasoning, intuition, tentative tries or contradicting approaches, remain of dominant importance.

Decision making process

Taking a process-oriented view of decision making, one can distinguish the following consecutive phases, according to Simon (1960):

— intelligence,
— design,
— choice, and
— implementation.

Intelligence has to do with the search in the decision context, i.e. looking up necessary data, isolating the raw material of the decision process and identifying more precisely the problem domain and, if possible, its structure. As already mentioned, one could also include the identification of the problem by one actor or the interaction process of a decision group in defining a problem. The latter might be the most delicate phase, and it is correlated with social processes and interactions.

In the *design* phase, different possible abstract models of the problem have to be created and analysed. As a consequence, specific courses of actions are correlated with these models. The models may also contribute to a better understanding of the problem domain. A primary feasibility analysis of possible solutions should be performed.

The selection of a plan from the set of possible alternatives is performed in the *choice* phase. This step may be a rather technical one, using methods of Operations Research, but it may also include some techniques for respecting subjective and intuitive judgement.

Implementation deals with the execution of a selected plan; it could also be regarded as part of the choice phase.

Fig. 1.3 also reflects the contributions of other information processing

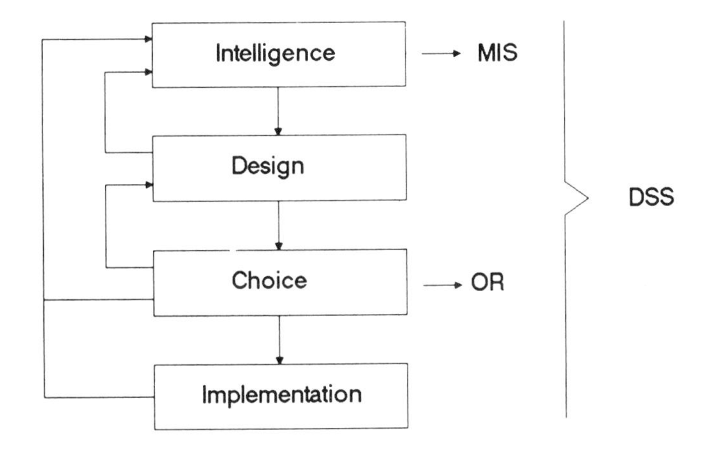

Fig. 1.3 — Phases of the decision making process.

tools to the decision making process. MIS can support the intelligence phase very well (under the assumption that in this phase problem definition is seen as a rather technical, solvable task). Operations Research techniques contribute to the choice phase. One possible way DSS can be used is in the integration of the single steps, with special emphasis on the Design and Choice phases, but also on the Intelligence phase. This may be obtained by the modelling capabilities of DSS.

Use of models

The integration and administration of mathematical models within a more general framework can be identified as the specific feature of the concept of DSS . The special power of DSS lies indeed in their capacity to perform data retrieval functions together with ad hoc analysis, including complex model building and execution. In a review of different systems, Alter (1977) distinguishes six different types of model invocations inside DSS. The typology consists of:

— retrieval of data,
— execution of ad hoc analysis,

— generation of standard reports,
— evaluation of consequences of proposed actions,
— proposal of decisions, and
— decision making.

Thus, typical models, which include such database management system functions as data queries and data manipulation, range from simple arithmetic functions and statistical operations to the ability to call up optimization and simulation models. In fact, the scope of a DSS lies in the integration of such different facilities. The idea of DSS, which is supported by the framework presented herein, integrates different fields of science, and not only such areas of computer science as database management, modelling techniques, decision theory, knowledge engineering and formal logics. It also puts weight on social circumstances by discussing the overall context which may determine or influence problem definitions and solution strategies. Thus, the concept and the construction of DSS call for an interdisciplinary approach. It should be noted, however, that interdisciplinarity has to be respected in every large system design, computerized or not, and that such systems cannot be seen as isolated entities.

Before describing architectural and constructional aspects of DSS, it is necessary to summarize their overall objectives, which may also be used in evaluating different systems and clarifying system boundaries.

General objectives of DSS

The problem domain, the decision which has to be made, is characterized by little structure. No attempt is made here to suggest a process which helps to identify or to clarify consciousness about a problem. Social interaction is responsible for this. But it may be hard to draw a border between problem identification, on the one hand, and problem structuring, on the other. The structuring or reformulation of a problem in an initial phase may also be connected with the winning of consciousness, but intuitively the difference between identification and structuring should be clear. Lack of structure may be due to novelty, missing data and knowledge, unknown context, etc. Although it is mainly the responsibility of the human user to structure the problem under consideration, there should be some machine-stored knowledge that supports this process. For example, possible approaches might be techniques for storing and retrieving information on similar problems that have already been solved, or automated question-answering procedures for clarifying items of the problem description that are not well defined.

The system should enable decisions to be made at different levels of an institution, be it an industrial, business or governmental one. In the general framework, there should be a distinction between techniques for the different levels, i.e. strategic, management and operational. These different levels might be represented by several software environments for users in different positions inside an organization. The distinction between these levels of decision making is based in part on the time frame for reaching solutions and on the complexity of necessary procedures. At the operational level, the model structures are well defined and the choices of the human user are limited; at the other levels, the importance of human judgement increases. At the operational level, the user must input some parameter and data values; at the other levels, he or she might also have to choose between different models or to create new models corresponding to the different aspects of a problem.

The different phases of the decision process, beginning with the intelligence, the design and the choice and ending with the implementation phase, have to be respected. Their interaction and recurrence should be allowed for. Furthermore, it should be possible to test the consequences of an action, for example by simulating a chosen scenario. In addition, intuitive solutions and trial-and-error procedures, which are close to the human way of decision making, should be supported by such a system.

DSS must therefore combine different basic techniques of computer science. These range from database management system functions, such as data queries or data manipulation, to the possibility of calling up mathematical models of various types and degrees of complexity (i.e. statistical procedures, optimization algorithms, simulation programs). Moreover, a decision maker should have techniques available on which he or she can build new models as the need or the context of a problem may change. So not only is the use of existing models of relevance, but also the possibility of fast and easy model prototyping.

A DSS should allow for the integration of non-normative models, i.e. the possibility of introducing personal subjective judgements which may be based on hidden knowledge, personal interest or emotion. This is closely related to the above-mentioned feature, namely that a decision maker is able to create new and personal models.

The system is under the control of the user. The human being is the major part of the symbiotic man–machine system. The user can choose between a variety of possibilities, and it should be possible for the user to create his or her personal environment. Finally, the decision maker should be free of technical burdens, but the system has to guarantee the transparency of the solution procedure used.

As a consequence of the features mentioned above, it is clear that a DSS has to be an adaptive system that allows the responses to new demands. The system has to be an open one, enabling integration of new parts and relevant facts. Thus, the use of an open approach for the construction of such a system can also be seen as one of the features of DSS.

1.5 ARCHITECTURE OF DSS

In the present section, a conceptual architecture will be developed which respects the objectives and properties outlined in the preceding sections. The architectural aspect of DSS may serve several purposes: to give a potential designer a conceptual tool for constructing a DSS; to support the evaluation and characterization of possible systems; and finally, to present a more practical and constructive definition of a DSS.

Two approaches, which seem to be representative of existing DSS, are discussed below.

A functional approach

Bonczek *et al.* (1981) take a decision process oriented point of view in their approach. Their central scope lies more in the single functions which a DSS has to fulfil during the process of supporting a decision. On the other hand,their concept seems to be general enough to allow the integration of further parts as the needs of the user or DSS designer change. The approach distinguishes three components, which differ with respect to their functions and not with respect to the software tools needed to satisfy the different tasks. The single parts are:

— the Language System,
— the Problem Processing System, and
— the Knowledge System.

This differentiation shows a similarity to the human decision making process: the formulation of the problem corresponds to the Language system and the elaboration of a problem to the Processing unit in connection with separately stored information in the Knowledge system.

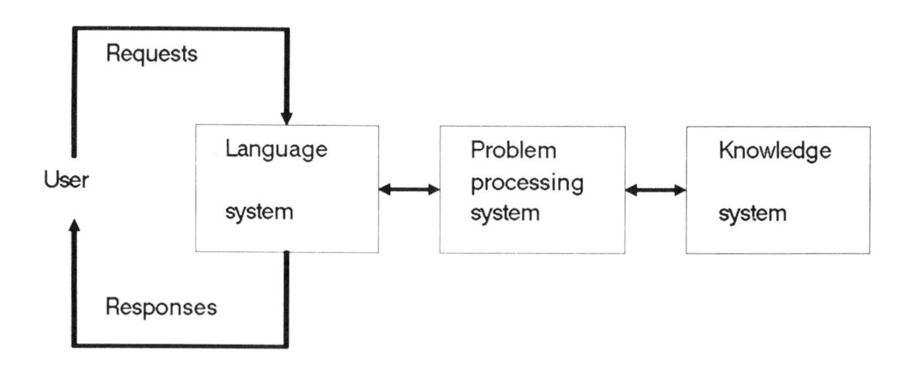

Fig. 1.4 — DSS architecture with respect to Bonczek *et al.* (1981).

This approach does not explicitly represent modelling or data retrieval functions. The user states a problem using the Language system, and the system responds by activating its Problem processing system and looking up specific information in the Knowledge system.

The Language system is characterized by the possibilities that it offers to the decision maker through the commands, instructions or actions that it allows the user to make. It is a vehicle that gives the user the possibility to express himself or herself to the system and reacts to specific situations, the channel by which the user can enter the system. Possible implemented techniques may be:

— Menus,
— Command languages,
— Fill-in forms, and
— Natural language interface.

The Knowledge system is the central memory of the system and contains knowledge about the problem domains. Since, in this approach, no distinction is drawn between the memorization techniques used, the following tools to structure information can be included in this component:

— Databases,
— Text,
— Rule sets,
— Procedural models,
— Frames,
— Spreadsheets,
— Vocabularies.

As one can see, different knowledge representation techniques are incorporated into this component. There are declarative ones, which store such information as rules and facts and which state what is true in the domain of discourse. And there are procedural ones, which describe how to solve a problem. At the level of an architectural description no clear distinction can be made between the single methods and their integration. Furthermore, static knowledge, normally stored in the form of rules, and time-dependent behaviour, reflected by dynamic models, is mixed up.

The Problem processing system seems to be the central and also the integral part of the architecture. It takes an input from the Language system, elaborates the query with respect to the information stored in the Knowledge system and finally returns the result to the Language system to respond to the user. This part of the system accepts specific tasks and handles them in cooperation with the Knowledge system. Furthermore, for every form of knowledge representation, some processing element has to exist in the processing system. For example, if there is a database in the Knowledge system, the database management functions have to be incorporated in it.

Problem processors may differ in their generality. On the one hand, there may be very specific procedures for handling particular problem situations. On the other hand, some general problem processing capability, probably implemented as an inference engine (i.e. problem solving procedure) in rule-based systems, may result in a form of a general problem solver. This component may also incorporate the different steps and levels of the process in decision making. Problem processing capabilities may be implemented in such different forms as:

— Database management,
— Text processing,
— Model execution,
— Inference engine and reasoning,
— Spreadsheet analysis, and
— Statistical analysis.

The proposal of the authors is so general that it also allows an Expert System to be included in the concept of a DSS. If one considers stored rules and facts to be a knowledge system and the Expert System's inference mechanism to be a Problem processing system, the equivalence between a DSS and an Expert System can be drawn. But this point of view conflicts with the requirement cited in the previous section that a DSS should integrate the use of a wide range of possible mathematical models, for example for optimization or simulation. Although they can be seen as tools for decision making, Expert Systems exhibit other characteristics lacking in

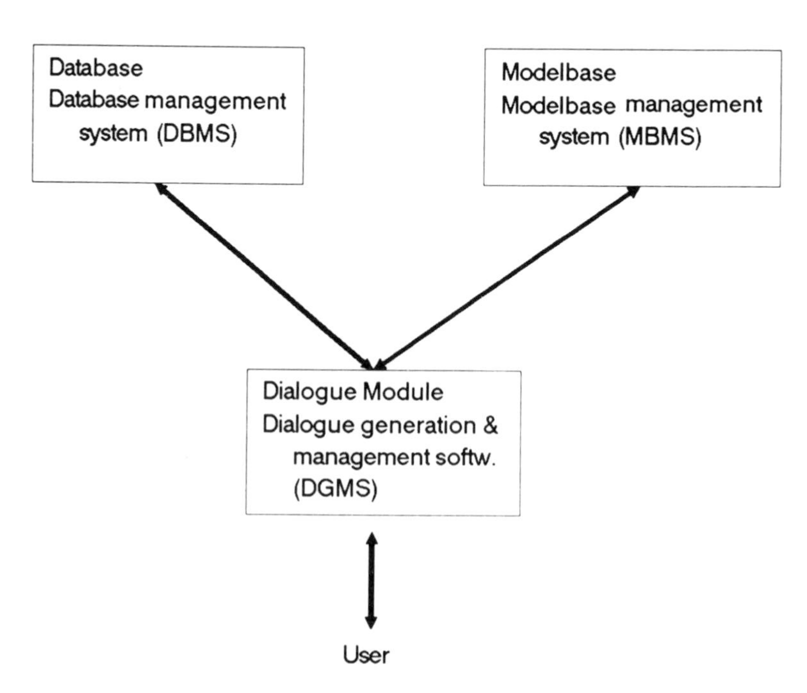

Fig. 1.5 — Architecture of DSS according to Sprague and Carlson (1982)†.
†The original design of Sprague and Carlson is slightly different. They add the DBMS and the MBMS to the Dialogue Module and call this last component the 'Sofware System'.

DSS and were developed for different scopes (see Chapter 6 and also Ford 1985). Thus, Expert Systems include implicitly the *substitution* for a human decision maker, whereas a DSS is built to give the decision maker advice and basic expertise to *support* the decision process. The possibility of integrating classical mathematical models with database management and also with Expert Systems techniques is a unique feature of DSS.

A tool-based architecture

The second approach, which also provides an overview of the wide range of conceptual possibilities, is the one taken by Sprague and Carlson (1982). Their point of view is correlated with the single technologies used to build a DSS. Thus, this architecture gives a potential DSS builder a more concrete concept in reviewing the necessary tools. Furthermore, Sprague and Carlson describe in more detail the capabilities of the different components, which are:

— the Database,
— the Modelbase, and
— the Dialogue Module.

The approach of Fig. 1.5 is less general than that of Fig. 1.4. The main components respect the data retrieval, the modelling, and the model invocation functions. Other aspects of the functional description by Bonczek *et al.*, such as specific problem processing tasks or knowledge representation capabilities, are not included or are hidden in the modelbase or the Dialogue generation and management software (DGMS). However, this concept seems to be clearer as it restricts the possible technological or basic tools.

The database component is of central importance in this concept of a DSS. This can also be derived from the fact that many DSS arise from some distinct database applications. In addition, the authors emphasize that this component also serves as an interface to foreign data sources and that it is able to include these data using data extraction procedures. The database management system (DBMS) may also serve to integrate the other components. For example, in the case of a sequence of several model invocations during a solution finding process, the database management system can manage the intermediate results which have to be passed between the single models. In such a case, the distinct models should possess unique operations to access the database. Some capabilities of the data subsystem, as listed by the authors, are:

— a combination of different data sources,
— data extraction,
— ability to add new data sources and to delete old ones,
— ability to handle personal or non-official data for personal judgments, and
— open interfaces to the other components.

The model subsystem also has the advantage of presenting a modelbase management system (MBMS) that serves a similar purpose as a DBMS in the case of a database. This subsystem is also of importance because it distinguishes DSS from other information processing tools through the incorporation of models and its modelling capability. The models included should respect the different levels of decision making (strategic, managerial and operational) and they should provide some support to the different steps of the decision process. The models also have to contain common features to be incorporated into a single base and to enable their integration. Such features might be common data access functions, by the database management system, or uniquely defined interfaces to pass information directly between them. Capabilities of the model subsystem include:

— the possibility of creating new models by prototyping,
— ability to integrate and to combine models to form compound models with a more complex model topology,
— ability to maintain a wide range of models for different levels of users,
— ability to interrelate models with appropriate linkages, and
— similar management functions as are provided by the database management system for a database.

The dialogue subsystem can be seen to fulfil the same functions as the Language system in the approach of Bonczek *et al.* (1981). But some additional possibilities are added by the supplementary insertion of a dialogue generation and management software. There might be overall control and system management functions which were included in the Problem processing system in the previous approach. Fig. 1.5 places some emphasis on the fact that the dialogue component might also serve for integration needs, for example in controlling the protocol of sequential model calls. The dialogue techniques used in this component are similar to those already mentioned. The following general capabilities are quoted by the authors:

— handling of a variety of dialogue styles to respect the personal needs of a user,
— supporting a variety of input devices,
— presenting output data with the help of different presentation techniques and devices, and
— performing input error checks.

Proposal for an extended architecture

In the following section, a conceptual architecture is developed which respects the objectives and properties discussed earlier and which also takes advantage of the possibility of combining the two architectures described above, i.e. a functional as well as a tool-based point of view. The proposal seems to satisfy the necessary functions of a DSS and provides at the same time constructional advice. In that sense, it is not contradictory to the approaches described above; rather, it is a slightly different view of a DSS.

The reader should understand this architecture, as well as the others, as a concept; it is possible that a concrete application of it may result in different designs at the implementation level (see Chapter 7). What is important is that specific functional blocks have to be separated and provided. In addition, the basic technical tools that exist to construct a DSS are described. Such a modular approach allows both the addition of new modules as the needs of the decision maker change as well as the integration of new tools when technological changes occur. Thus, it is an open concept, enabling adaption and evolution.

In Fig. 1.6, an architecture of a DSS is presented which is a combination

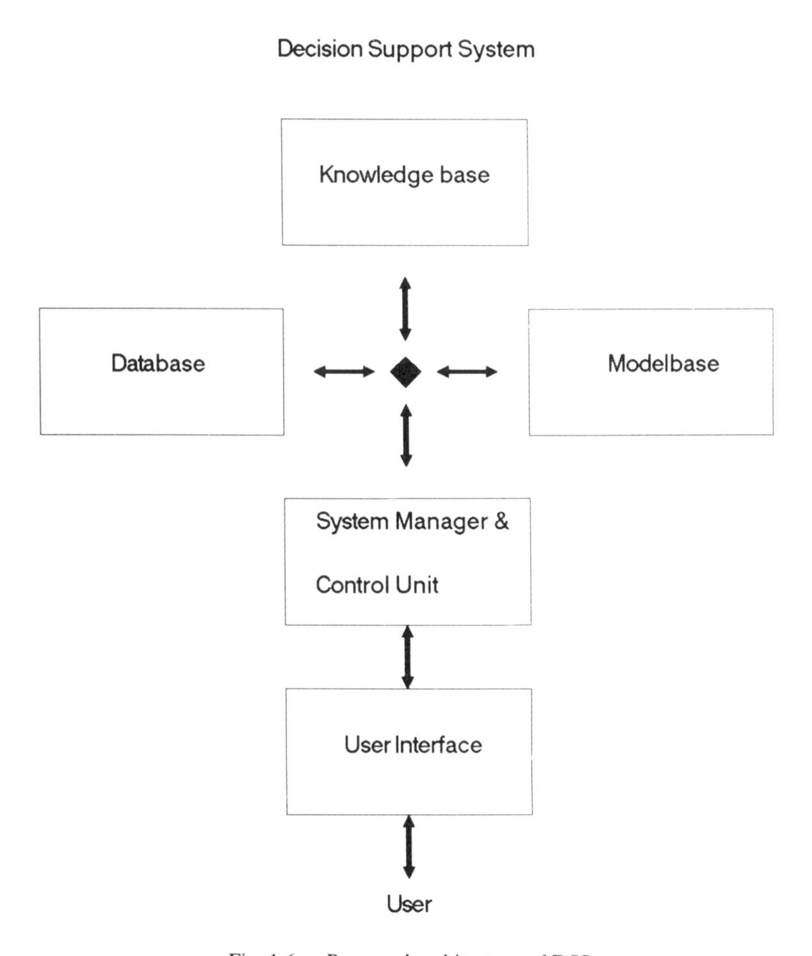

Fig. 1.6 — Proposed architecture of DSS.

of the two described above. It adds a Knowledge base with respect to the definition of Sprague and Carlson (1982) and splits up the components provided by Bonczek *et al.* (1981). It explicitly includes a Modelbase, based on the requirements that a DSS has to contain procedural models, and thus it is clearly distinct from an Expert System. In that sense the proposed architecture is more specific than that of Fig. 1.4. On the other hand, by distinguishing between the different bases which have some parallels in the field of computer science, the approach seems to be a modular and adaptive one. As already mentioned, it allows for not only the easy integration of the different subsystems but also the adaptive enrichment of the distinct bases. The links in Fig. 1.6 are tentative ones, leaving the specific integration and construction open.

The user interface (Dialogue module) is the part which is closest to the architectures already discussed. Its function remains the same, as it constitutes the unique interface between the user and the system. One difference might be that specific rules or knowledge, which guide the interaction between user and machine, may be kept in the knowledge base. This component may store, for example, recent usage habits and offer, with respect to these data, a modified interface in subsequent sessions to support a specific user. So a help screen could be called automatically each time the user attempts a command which failed the last time.

The Database is explicitly mentioned as a specific information and knowledge representation tool because of its main importance for DSS. Most applications are highly data intensive, and historically many DSS are successors of database applications. The Database has to handle different types of data with their logical relationships, time-series and scalars, both experimental and measured ones (which means that there are different degrees of confidence, i.e. hard and soft data), both external and internal data. And, as already mentioned, in a distinct implementation, the Database may serve as an integration tool for the whole system, owing to not only its capability to store data but also its well-developed database management system functions.

The content of the Modelbase can be seen as the procedural representation of knowledge, in contrast to the Knowledge base, where information is stored in a declarative form. This component reflects the time-dependent behaviour of real systems and constitutes the main distinction between DSS and Expert Systems. In the latter case a knowledge base together with an inference engine are not well suited to describing the dynamic time dimension of real world entities (see Chapter 5).

Different types of models for the various steps of decision making previously mentioned have to be maintained. The base will contain both stand-alone models and such models, also called submodels, which have to be connected to be applicable to solving certain specific problems. Furthermore, for handling new problems or for creating a personal modelbase which respects the specific needs of a decision maker, the possibility of creating new models, i.e. the availability of model prototyping, should exist.

The possibility of using qualitative modelling procedures would be another positive feature. Such procedures allow the user to describe real problems in qualitative terms and to watch the qualitative behaviour of a system, (e.g. variable 1 is always greater than variable 2, or the cause–effect relationship between different entities; see also Chapter 5 and Forbus 1984). This would allow the user to integrate some reasoning about time and would also give the system the power to support tentative as well as intuitive approaches. In addition, it should be possible to import externally developed models or programs. This enables the evolution of a DSS with respect to new technologies. Some technical conditions seem to be necessary for satisfying these different objectives such as a unique model invocation possibility, similar model interfaces or an abstract representation of the models which respects some common constraints.

The Knowledge base is a supplementary element with respect to the tool-based architecture, and it has a more specific scope than in the functional approach. It is designed as a distinct base for segregating the knowledge necessary for the functioning of the system in cooperation with the other components and it also includes specific problem information. Knowledge may be represented by using well-known techniques such as first-order logic or frames (Minsky 1981). Knowledge which is specific for the functioning of the system may contain the abstract and common form of model representation, the rules for constructing models, or constraints for guiding the user interface. In that sense, the Knowledge base also has integrating tasks inside a DSS. Although there has been some research on integrated database technology with the knowledge representation mechanism, resulting in so-called knowledge base management systems (for an overview, see Brodie and Mylopoulos 1987), a distinction has been made here between the Database and the Knowledge base, as they serve different purposes in the present context, although both store information.

The System manager has mainly control purposes. Some of its functions can be kept in the other bases, but it would be preferable if the storage of data, models and knowledge were not mixed up with the dynamics of their use. For example, it could be argued whether the modelbase management system should stay inside the Modelbase or be included in the System manager software, but this seems to be a more specific system building question. In the present representation, it is better to distinguish this block from the others in order to maintain conceptual simplicity. For example, the ability to construct compound models by connecting basic models of the Modelbase with the support of specific knowledge in the Knowledge base could be a possible function of this unit. Still another function could be keeping control in the case of a sequence of several model calls.

This component plays an essential role in integrating the other parts. It is thus located near the centre of the system, and it shows features that are similar to those in the Problem processing system in Fig. 1.4.

REFERENCES

Alter, S.L. (1977) A taxonomy of Decision Support Systems. *Sloan Management Review* **19,** No. 1, 39–56.

Anthony, R.N. (1965) *Planning and Control Systems: A Framework for Analysis.* Harvard University Graduate School of Business Administration, Harvard, Massachusetts.

Bell, D. (1973) *The Coming of Post-Industrial Society.* Campus, New York.

Bonczek, R.H., Holsapple, C.W. & Whinston, A.B. (1981) *Foundations of Decision Support Systems.* Academic Press, Orlando, Florida.

Brodie, M.L. & Mylopoulos, J. (eds) (1987) *On Knowledge Base Management Systems: Integrating Artificial Intelligence and Database Technology.* Springer, Berlin.

Forbus, K.D. (1984) Qualitative process theory. *Artificial Intelligence* **24,** No. 1-3, 85–168.

Ford, F.N.(1985) Decision Support Systems and Expert Systems: A comparison. *Information and Management* **8**, 21–26.

Goldstine, H.H. (1971) *A History of Numerical Analysis from the 16th through the 19th Century.* Springer, New York.

Head, R. (1967) Management Information Systems: A critical appraisal. *Datamation* **13**, No. 5, 22–28.

IEEE Spectrum (1987) How computers helped to stampede the stock market. *IEEE Spectrum* **24,** No. 12, 30–33.

Keen, P.G.W. (1987) Decision Support Systems: The next decade. *Decision Support Systems* **3,** No. 3, 253–265.

Landes, D.S. (1968) *The Unbound Prometheus.* Cambridge.

Landry, M., Pascot, D. & Briolat, D. (1985) Can DSS evolve without changing our view of the concept of 'Problem'? *Decision Support Systems* **1,** No. 1, 25–36.

Lewandowski, A. & Wierzbicki, A.P. (1987) Interactive Decision Support Systems — The case of discrete alternatives for committee decision making. *Working Paper 87-38,* IIASA, Austria.

Minsky, M. (1981) A framework for representing knowledge. In: Haugeland, J. (ed.), *Mind Design.* MIT Press, Cambridge, Massachusetts, pp. 95–128.

Parker, B.J. & Al-Utabi, G.A. (1986) Decision Support Systems: The reality that seems to be too hard to accept? *OMEGA Int. J. Management Science* **14,** No. 2.

Schwendter, R. (1982) *Zur Geschichte der Zukunft. Zukunftsforschung und Sozialismus.* Vol. 1. Syndikat Verlag, Frankfurt/M, FRG.

Simon, H. (1960) *The New Science of Management Decision.* Harper & Row, New York.

Sprague Jr., R.H. & Carlson, E.D. (1982) *Building Effective Decision Support Systems.* Prentice-Hall, Englewood Cliffs, New Jersey.

2

Computer systems for environmental problems

Concluding one of his major studies on the planning of the River Nile, probably one of the largest environmental problems ever tackled in the world, H. Hurst acknowledged the fundamental work performed by 'computers' (Hurst 1952). At that time, in fact, the Egyptian Ministry of Irrigation, ruled by the British, had a 'computer room' crowded with a large number of clerks performing all the calculations needed to manage and to plan such a wide river basin.

A few years later, in 1958, an 'electronic computer' was used on the same problem in order to find the optimal allocation of water between Egypt and the Sudan in a study for the Sudanese Ministry of Irrigation and Hydroelectric Power (Morrice 1958). At that time, the water system contained five major dams, and several other barrages and control points. Many new structures had been proposed, of which the most important was the future Aswan High Dam. Monthly volumes of inflow in the period 1905–1942 were used as input data to a model describing the system in terms of mass balance equations at all the major junctions of the river and control equations at 16 control points. The system was developed on an IBM 650 which allowed the effect of 300 different project combinations to be tested, thus becoming probably the first real example of a computer-assisted environmental plan. The Nile became a showcase for the efficiency of using electronic computers in this kind of study. It was estimated that a 50-minute computer run could replace 1500 man-hours of traditional work!

Still in the sixties, in the first modern work on the managing and planning of water resources, which laid the basis for most environmental models in the following years, Maass *et al.* (1962) referred to the computer as an 'electronic computer' in order to distinguish it from accountants. However, the use of such devices was spreading rapidly and Chow (1964) was able to quote 51 papers reporting its application in the field of hydrology alone.

Today the meaning of the word 'computer' has been definitely established, in environmental as in other sciences, even if a lot of work has still to be done to permit the effective and generalized use of this tool.

If one goes back to the pioneering work on the applications of computers to environmental problems (see, for instance, Chow 1964), one realizes that their spread has been paralleled, if not induced, by the development of mathematical formulations of environmental phenomena. During the sixties, in fact, the basic role of computers was in performing calculations, and as such they were used also in the environmental sector. Although the mathematical expression of many physical laws had been known for many years (sometimes for centuries), it was not until that time that the traditional notion of 'formula' was replaced by the more modern term 'mathematical model'. The latter is usually a collection of algebraic or differential equations which tries to represent not only a single phenomenon (for instance the diffusion of a pollutant in the atmosphere) but also the entire complexity of the causes and effects of that phenomenon (for example the economy of the processes generating the pollutant, its combustion and emission dynamics, and the damages caused by its diffusion).

While the first approach (physical formulas) was in various degrees amenable to an analytical study of some very simple conditions (diffusion among a homogeneous quiescent fluid), the mathematical models under development were too complex to be analysed by hand and were driven by a substantial amount of data concerning real conditions in order to allow the solution of practical problems. They necessitated an unprecedented amount of computing power, which could be found, even if at a relatively high cost, in the new computers with which several research centres and some environmental agencies were equipped (sometimes for the basic purpose of performing accounting).

The well-known trends in prices and performance of computer equipment have led to its diffusion to almost all human activities and thus also to environmental institutions. This penetration (and that of mathematical models) has not been even in all the different branches. As an example, several water resources studies were based on the use of computers long before the first attempts were made to solve an atmospheric pollution or population dynamics problem with such a device. This is certainly not due to intrinsic differences in the problems dealt with but probably to the different backgrounds of researchers involved in the studies. Water resources problems have usually been studied by people with an engineering background, who probably felt more comfortable with electronic machines and the formal and precise codes that were needed to operate them. Today these differences have become less perceptible although they still exist, as will become apparent to the reader in the following chapters of this book.

Technological changes in computer hardware and software and psychological changes in the users have induced, particularly in recent years, a major shift in the environmental applications of computers. The computer is now used not merely to perform calculations but also to collect, store and retrieve data, to present them in different ways, and to help people in understanding them. In other words, the computer is no longer an isolated and protected device in an inaccessible room, but has become a flexible, multi-purpose tool for supporting all activities of an environmental engineer: a complete 'information processor', part of a complex information system. This system normally contains, in addition to the computer(s) and the operator(s), a measuring and transmission network and an implementation network, as in many other process control systems. Through this system, the computer may collect, store and retrieve environmental information (digital values, analogue signals, images); it can elaborate it and transmit results to different locations, people and instruments. Furthermore, the development of software engineering and particularly of user–machine interfaces has eased the access of a much larger number of people to the use of computers, thus attenuating the fear, that was fairly widespread until the beginning of the eighties and the advent of personal computing, that these machines may overtake their own developers and become the real actors in any human decision.

Once equipped with sufficient software (or, to use the term introduced in the previous chapter, a modelbase), a computer may represent for the environmental scientist the equivalent of the traditional physics or chemistry laboratory for those concerned with these two disciplines. Without using a computer, it is in fact impossible to experiment on environmental problems since costs, risks and time to make tests on real systems would be unacceptably high. In the same way as an experiment in a chemical laboratory requires some careful preparation with regard to both the problem and the equipment, some knowledge about the behaviour of the environment and of the experimental device (the computer) is necessary for the environmental scientist to be exactly in the same position as his or her colleagues. There is no need to prove that the idea of experimenting may be useful in science (while measuring the fall of an object from the Tower of Pisa, Galileo could probably never imagine that years later an experiment on subatomic particles would need a laboratory of hundreds of square kilometres and billions of dollars of investment). This is a well-established idea since 'one learns from errors' and experimenting with a computer allows the environmental planner or manager to make mistakes without being dismissed.

Starting from these bases, the computer may enter even more deeply into the solution of many environmental problems. It may carry out all of the activities that do not require a critical review by the decision maker, it may test different alternatives, show their drawbacks, review past experience, and suggest choices. It may supply all of the information needed for a more conscious decision on many environmental problems. The achievement of this goal requires the development of software environments which allow easy manipulation of mathematical models and environmental data and integrate these models and data with some techniques developed in the field of artificial intelligence. Finally, this software must have an interface to the user that could free him or her from all traditional computer handling problems (using operating systems, compilers, files, etc.) which have, on many occasions, prevented the extensive use of these machines.

Before applying the DSS architecture outlined in Chapter 1 to environmental problems, this chapter will review some known methods of, and open questions in, applying models and computers to this area and clarify what is meant by a (computer-based) environmental decision support system. Furthermore, a typical user of the system will be described, as well as the situation in which the computer operates. The chapter concludes with a series of real examples of environmental applications, which show the different roles the computer can play.

2.1 DEVELOPING MODELS FOR ENVIRONMENTAL PROBLEMS

The modelbase plays an essential role in the architecture of any DSS. The models it may contain are usually divided into descriptive and decision (sometimes called normative) models.

The former simply portray the behaviour of a certain physical system as a consequence of external inputs (also called forcing factors or driving forces). All models used to simulate or to forecast the development of natural or human-induced phenomena fall into this category.

Decision models are normally aimed at suggesting a decision on how to act on a physical system in order to let it perform in a desired way. Planning and management (or control) models are of this type; the main conceptual difference between them is that, in the case of planning, the decision is taken only once for a long period of time, while management implies repetitive decisions which are close enough in time to influence each other. As pointed out in section 1.4, this time range usually corresponds to the degree of structure of the problem. Planning is much more likely to involve several unclear elements, which are less relevant for short-term decisions.

The differences between these various types of models and the way they may be coordinated into a modelbase will be discussed in Chapter 4. In the present context, however, it should be stressed that, whatever mathematical model one is trying to develop, in principle all the steps represented in Fig. 2.1 should be accomplished.

Problem perception

The importance of this first step was dealt with in the previous chapter. It must be recalled that problem perception is largely subjective, and thus two people confronted with the same situation may conceive of two quite different formal ways to deal with it. This step, while being part of model development, should be directly performed by the decision maker, i.e. by the person or institution responsible for taking the final decision on implementing the solution of the problem. Too often in the past, the analyst has taken the responsibility for defining the problem, and sometimes he or she was biased by the availability of solution methods more than by the core of the problem itself. This is probably one of the causes of the weak practical impact that environmental modelling has had in comparison with the massive research effort in the area. Strict cooperation is needed between the decision maker and the analyst to produce an effective and viable problem identification.

Conceptualization

This stage represents the formulation of the problem in analytical terms. It should be noted that 'analytical' in this context does not mean a procedural

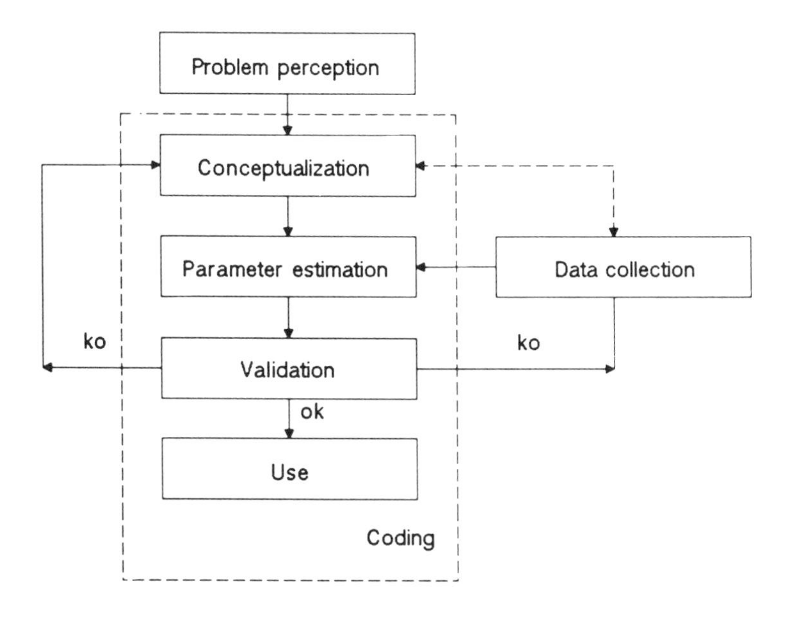

Fig. 2.1 — Flowchart of model development.

formulation, i.e. a set of mathematical tools which allow the problem to be solved; rather, it may well be constituted by a set of 'rules' that express in descriptive terms what is known about the problem. In most cases, however, the output of conceptualization is a set of equations which should capture the mechanism of reality. This output is not at all unique, and the same problem can often be formalized in several different ways. Again, the choice between different models is largely subjective and this is the reason why, for building a practical instrument to support decisions, a whole range of models should be made available. This justifies once more the choice of including a modelbase in the architecture outlined in the previous chapter.

A last, but important, point about conceptualization is the lack of definite methods to perform it. The ability to formulate the problem in a correct way usually derives from the experience of the analyst, which also means from a critical appraisal of what has already been developed in the field.

Data collection

The formulation of the model should in principle follow a certain analysis of the data and influence the subsequent more accurate data collection. The first analysis is in many cases redundant since the basic mechanisms of a number of environmental phenomena are known, but it may be essential when formulating a model for an entirely new problem. Usually, the first analysis consists of statistical tests (several software packages are available for any type of computer), particularly tests which may highlight the causal link between some variables (see, for instance, Finzi *et al.* 1988).

The fact that further data collection must depend upon the model has often been disregarded in the past. It has been very common in environmental agencies to start collecting data (particularly after cheap computer mass storage became available) in the hope of creating a 'universal' data bank — an archive capable of containing any information needed to answer any question. However, one often realizes that the data necessary to implement a given model are missing, since model input and output data must be available in a precise sequence to be of some use, and they are almost always collected independently.

Parameter estimation

This is normally a very technical step in which the model that has been formalized is fitted to the particular situation under study. This means that the parameters of the equations (or of the rules) are estimated in order to represent as close as possible the data available on the case at hand. The parameters are normally estimated (the terms 'calibrated' or 'tuned' are used by some authors) in two different ways, depending upon what they represent. If they correspond to some rate or relation which can be measured on the real system, they are just the results of those measurements or are computed through some experiments. In many cases, however, it is too expensive or risky or time consuming to measure the relevant values on the real system; or the parameters represent a relation which has simply been hypothesized or the aggregation of several effects. These parameters are thus estimated by solving a mathematical programming problem that minimizes a measure of the differences between real data and those obtained from the model. As an example, consider the fate of toxic substances dispersed into the environment and passing from one organism to another until they end up in harmful concentrations at the top end of the food chain. It is impossible to follow their accumulation within a living species, since the only feasible experiment is to measure their concentration in a dead organism. Furthermore, the precise mechanisms of concentration and release of these substances from organisms are largely unknown. One possibility is to use a simple bioconcentration model, with only two compartments: the environment and the organism. Starting with data on the substance concentration in the organism and in the environment at different times, it is possible to estimate the uptake and release parameters (which determine the exchange of substance between the two compartments) using a method that minimizes the simulation error. It is even more important, in cases like this, to compute the variances of estimated parameters; this allows to assess the statistical significance of the estimates obtained from the available data (see Galassi *et al.* 1988 for details).

It is interesting to note that in almost all cases the model error is measured by the well-known sum of the squares of the differences between model results and real data, which has come to be known as 'least squares' parameter estimation. However, a careful analysis of the problem to be solved may sometimes suggest other criteria for tuning the parameters. Since a model always constitutes an approximation to reality, it would be wise to try to get better performances (i.e. be closer to actual data) in those particular circumstances which appear to be most influential in solving the problem. One may consider, for instance, estimating parameters just by looking at extreme episodes or at some other particular condition.

Validation

Before using the model to take real decisions, it is important to check whether it is robust enough, i.e. whether its capability of mimicking the real system is not confined to the set of data which has been used to formalize and to calibrate it. If the model can be proved to perform satisfactorily with different sets of data, it has a better chance of having captured the important mechanism of the real system. It is thus necessary to have additional data and let the model work with them. Again, what 'satisfactory' means is a matter of debate: by definition, each researcher on environmental modelling means the performance of 'his' or 'her' model, whatever statistical results it has. Also, there is no specific method to carry out validation, but again one should be guided by the type of problem which must be solved. In many environmental problems, for instance, the performance of a system under critical conditions is of more concern than its behaviour under normal circumstances; thus, the set of data used for validation must contain significant episodes of this type. A model which can always guarantee a maximum deviation from reality is often more acceptable than a model which is almost always quite precise but which may deviate in critical cases in an unbounded way. This is due to the natural risk aversion of all decision makers, who prefer to be guaranteed against very critical situations in exactly the same way as anyone who pays a small insurance premium every year (maybe for an entire lifetime) to protect himself or herself against some improbable, but potentially catastrophic, accident.

An unsatisfactory validation may be due to two main reasons. The formalization of the problem may have been poor, which means that the analyst has to change his or her mind and perhaps try to get more information from the decision maker in order to modify the description of the real system he or she has conceived. Alternatively, the parameters may not have been estimated with sufficient accuracy. This normally means that more data (or more accurate data) must be collected and the calibration has to be repeated. Sometimes the criterion for calibration may be changed to improve the performances that the validation has shown to be unacceptable.

Use

Only after a model has been validated can it be put to use. The usage of a model means, in the present case, to test on it the effects of one or more alternative situations, either to better understand the structure and the behaviour of the real system or to find how to act on this so that it behaves in a more acceptable way. It is essential to remember that the use must be consistent with all the previous steps, and thus a model should not be used to solve a problem for which it was not conceived. This is a common error made by people who think, for instance, that a given model of air pollution may be used for both short- and long-term pollution forecasts, although the atmospheric components involved in the two cases are completely different.

Coding the model development procedure

Programming (or using some software package) is normally necessary in various steps of the procedure outlined above. First, the conceptualized model is usually translated into a computer program; then it may be run several times as a subroutine of some optimization procedure for estimating the parameters; and, finally, it may be rewritten in a different version for actual use. During the entire model development procedure, several approximations are introduced (see Fig. 2.2). Besides the approximations introduced in the conceptualization phase and those due to the finite length of the computer words, other errors are due to the methods used to translate the model into an executable program. For instance, a continuous variable must be sampled in a finite number of time intervals, and an improper sampling may cause divergence problems in the integration of differential equations. The same is true for all optimization methods, which often consider some very low but non-null values as zeros. These kinds of problems will not be discussed here; they are widely treated in the literature (the reader may refer to any textbook on numerical analysis) and their effects on the solution can usually be established fairly accurately.

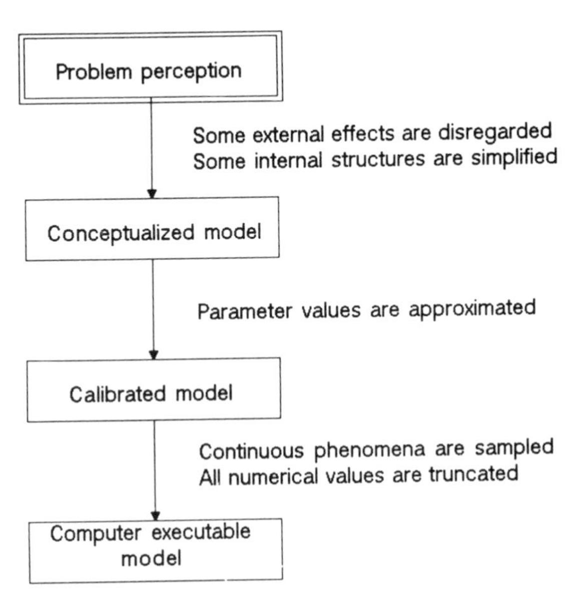

Fig. 2.2 — Errors in the model development procedure.

There is another implementation problem due to the use of computers. Since the estimation procedure may be very time consuming, it sometimes uses a simplified version of the final program and may require a larger computer and sometimes even a different computer language. This introduces an additional and often not quite predictable noise in the procedure outlined above, since the parameters may have slightly different values and the computer and the coding itself may cause differences in the way in which the various mathematical expressions are evaluated.

There is no guarantee that the computer code finally released for actual use may still be sufficiently precise to help in solving the original problem. It is thus important to perform the validation on the final version of the program on the target computer, since this is the only way to check whether the entire software/hardware system is robust enough to assist in making real decisions.

From this point on, the word 'model' will be used in referring to a validated set of mathematical expressions coded in any computer language.

2.2 CRITERIA FOR COMPUTER APPLICATIONS TO ENVIRONMENTAL PROBLEMS

Many recent environmental symposia and conferences have been held with the declared aim of filling the gap between computer experts and modellers, on one side, and practitioners on the other (see, for instance, Tavares and Da Silva 1986, Zannetti 1986, IAHS 1989). This is a clear symptom that there are still problems in the successful application of systems and computer science techniques to environmental planning and management. The large effort on the research side has not generated a comparable diffusion on the application side because certain criteria for applying computers in this area have not been completely satisfied.

Theoretical research has been going on to define abstract criteria for assessing the quality and the suitability of computer models. Ören (1984) quotes more than 80 papers on this topic. Henize (1984) and Meadows and Robinson (1985) have concentrated on quality criteria for models that are used to make social decisions, among which are environmental problems. Some basic requirements that should be fulfilled to make models easily accepted appear to be the following:

— accuracy,
— robustness,
— simplicity,
— transparency, and
— adequacy.

Accuracy

As already anticipated, accuracy is normally the major concern in model development. It means that the model represents reality in a fairly close way, and consequently decisions suggested by using the model can be implemented (almost) without any modification. A number of methods exist to check this characteristic of the model (Sargent 1984). Scientists seem to prefer 'objective' measures as statistical tests (even if the degree to which a test can be considered as satisfied is still a subjective judgement), while people involved in practical decisions prefer a direct comparison of model output and real data, both in terms of numbers and graphically displayed on the computer screen. In certain particular cases, the perception of small deviations, which cannot appear in any statistical analysis, in the model results may be the most important element in judging the accuracy of the model in practice.

Robustness

Robustness is the capacity of the model to filter out noise due to factors which were disregarded in the model formulation phase. Any formulation, in fact, only partially translates the complexity of the original problem, and several aspects must be simplified or ignored. In the same way the values used for parameters in the model simply represent an estimate of the true parameter values. The difference is due to the limited amount of data on which calibration is performed and to the method used.

When data differ from those used for calibration (and thus parameters differ from their 'nominal' values) and some unforeseen inputs enter the system, the accuracy of the model should not decrease significantly.

Simplicity

The model must be 'simple'; that is, the number of variables and parameters must be limited and decision makers must be able to understand its behaviour easily. The number of parameters is limited not only by the speed of execution, but also, in several cases, by the availability of data, since at least few tens of data for each parameter are required to obtain a reliable estimate. Furthermore, parameter estimation methods may provide poor results when confronted with very complex models, and one often finds that an improvement in the complexity of a model does not generate a comparable improvement in results.

However, the simplicity of the model itself is important in allowing people who are not familiar with these techniques to understand what the model means and why it behaves in a certain way even if this may contradict intuition or past belief. The lack of this characteristic prevents practitioners from having confidence in the model and thus from implementing the suggested conclusions.

Transparency

However, simplicity itself is not sufficient if it is not translated into an adequate computer code. This should allow the decision maker to look at single parts, variables or parameters or relations in the model and to modify them. In this way the user may 'play' with the model as much as he or she likes and may explore all the aspects of its behaviour. The possibility of modification enables more extensive testing of the model by the user and allows him or her to take advantage of the experience gained during model use. Sometimes this feature can be partially embedded into the model code by using recursive algorithms for parameter estimation or by using some artificial intelligence technique to improve or enlarge the fact base from which conclusions are derived.

Finally, the possibility of modifying the model has a clear psychological motivation. Users must perceive that they themselves are making the decision and that their responsibility is not taken over by the machine. The fear that a computer may prove to be more effective has motivated many managers to disregard it as an aid in their more difficult tasks. Computers are thus often used as data repositories or local controllers, but are rarely involved in the main decisions of environmental agencies. It is important to let users perceive them as tools that are able to perform more sophisticated calculations than old slide rules, but nothing more. There is no competition between the human and the computerized decision; the latter is simply the result of hypothetical well-defined conditions, which always differ from real ones, and as such is only an additional piece of information on which the final human decision is based.

Transparency is sometimes considered the most important characteristic of computer models and their major advantage over 'mental' models, those that anyone can use to make decisions. Even if transparent computer models are wrong, at least one can rigorously follow the path which leads them to a particular conclusion. One may suggest that this is typical of any computer program, but it is well known that logical errors may be so well hidden in a program that only a specialist can track them down.

Adequacy

The preceding features cannot be perceived, and thus have almost no practical meaning, if they do not emerge from the code in a form which is meaningful to the user. The suitability of the interface thus becomes a condition *sine qua non* for the acceptability of any model. The code should be able to speak the same language as the user, and the man–machine dialogue must be based on information which is known and accepted by the user. Chapter 6 will deal in detail with some of the problems connected with this aspect; at this point, however, it should be stressed that there are several ways to accomplish this. One is to target the software to a precise category of users. This means that a precise 'model' of user is known to the software developer and the required type of dialogue is included in the software specifications. The second is to prepare an interface which can interact according to various degrees of expertise of the users. Obviously, this means an increase in the time and cost required to develop software but the result may serve several different users within an institution. The choice of the level of the dialogue may be made directly by the person who sits at the keyboard or it may be made by the software itself, using some artificial intelligence features to test user understanding (see Chapter 6, and also Ushold *et al.* 1984 and Bundy 1985).

All of the criteria outlined above can only be judged from a subjective point of view, and therefore it is of great importance that they are agreed upon between the software developer and the program user. Sargent (1984) ranks this recommendation first in a list of possible measures to improve simulation software validity. Without a common definition of the way a software product should perform, the chances that it will never be used for any significant decision increase dramatically.

2.3 ENVIRONMENTAL DECISION SUPPORT SYSTEMS

The rest of this book is devoted to the analysis of the various facets of what will be called an environmental decision support system (EDSS), i.e. a computer system to help decision makers in environmental agencies and organizations. Thus a detailed analysis of all its major components will be carried out in the following chapters. It should be pointed out from the very beginning, however, that an EDSS also fits nicely into the general framework presented in Chapter 1, and thus it has no peculiar architectural features. However, the contents of its data, model and knowledge bases must reflect the features of this application domain.

Before turning to a detailed discussion of EDSS, it will be necessary to analyse the kind of problems which could be solved with its support and the human environment in which the authors believe the EDSS will operate. It is clear that the definition of EDSS that will finally emerge from this chapter and, probably, from the entire book is too vague to represent a paradigm to follow in the realization of new EDSS. Nonetheless, attention will be drawn both to some common problems and to some possible solutions useful in the realization of effective systems. Any further objective is outside the scope of this book, given the complexity and variety of environmental problems and the general debate still taking place on what a DSS should be (see, for instance, Keen 1987).

Some common features characterize environmental problems even if they are not unique to this area.

Dynamics

Environmental problems have a significant dynamic component, which means that the conditions of the real system, at the time the decision is made, are the results of all the past history of the system and influence its subsequent behaviour. This dynamic characteristic of the environment may be represented explicitly in the problem formulation or only implicitly by suitably formulating the problem constraints (see Chapter 4). Usually, however, this characteristic is translated into the model formulation by using a set of difference or differential (total or partial) equations.

This dynamic aspect is possibly the major feature of environmental problems and stresses the importance of having in the model base the possibility of accurately portraying it. In this connection, it is important to underline the role of simulation models, in which this characteristic can be more easily and intuitively represented.

Spatial coverage

Another essential aspect of environmental problems is their spatial dimension. While in many other application domains the problems under study are confined within very precise (and usually small) borders, for instance a plant, a reactor or a firm, environmental problems deal with spatially varying phenomena with no or very unclear borders. For example, circulation in the atmosphere is a typically unconfined three-dimensional problem, and the circulation of a pollutant in a reservoir takes place in three dimensions, even if the borders are more precise (it is still unclear how to determine exactly where the reservoir starts and the tributary river ends). This feature requires in principle the use of partial differential equations to model both the time and the spatial dimensions, but, in addition to their numerical complexity, it is often difficult to set sound boundary conditions for them. This is why a number of methods have been developed to lump spatially varying values into one or few aggregate measures. Partial differential equations reduce in this way to total differential equations, a tool which is less difficult to handle from a numerical point of view. Furthermore, continuous variables are always sampled at finite time intervals, and thus difference equations represent quite a common formulation for many environmental problems. It should be pointed out that this series of approximations may not result in a reduction in the accuracy of the model with respect to more sophisticated ones. It is, in fact, useless to utilize accurate mathematical descriptions, like partial differential equations, if the uncertainty about other information, for instance boundary conditions, is very high. A more crude (and simple) technique may have less technical problems and thus lead to results with similar (if not higher) accuracy.

The sectors of environmental science where spatial variability can be disregarded or where simplifying assumptions (such as spatial homogeneity) can be adequately made have benefited by the possibility of formulating problems in a more convenient way. Researchers in water resources, for instance, have developed a number of lumped models for superficial flows, while studies of atmospheric pollution have been based more on the use of two or three spatial dimensions.

The question on where to fix the model boundaries is also linked to the spatial characterization of environmental problems. Every model must in fact include a number of factors or a number of social entities or a certain area, but lump the effects of all the rest of the world into few input or parameter values. A careful decision on which part of reality can be formalized and which should be simplified is important in providing really useful suggestions to the decision maker. A support in this area may be, for instance, the possibility of running different models at a larger and at a smaller scale and then comparing their results. If they do not diverge too much, either may be used for further investigation. This again means that the modelbase must be rich in models with different coverage to allow these comparative evaluations.

Periodicity

Another important feature is periodicity. Although, in fact, long-term trends are sometimes important, even in environmental management and planning, in the great majority of the problems there is a strong periodic component due to the annual cycle of nature.

In some problems, cycles with a longer period appear: for instance, hydrological variables seem to be influenced by sunspots (which have a period of about eleven years) or salmon come back to their birthplace to reproduce (and can thus be caught) three to five years after having left it to reach the ocean. In other situations, a shorter periodicity can be detected: the chlorophyll photosynthesis of vegetation typically has a daily cycle due to sunlight and, again, for solar radiation, the temperature of the air close to the ground oscillates in the same way.

Human activities also have the same kind of periodicities (daily and yearly cycles), but may also induce effects with a new and 'synthetic' time scale. The release of a reservoir to produce hydroelectric energy may show a weekly cycle due to the decrease of power demand during weekends. Thus, a river flow which receives water discharged from that reservoir may also oscillate with a weekly component in addition to the normal annual cycle.

Determining the appropriate time scale of the problem at hand is of major importance in formulating it in mathematical terms and deciding correctly upon it. The periodicity of the phenomenon is an important factor in determining the management or the planning horizon for which a decision must be made.

Randomness

A stochastic component is always superimposed on the cyclic one in environmental problems. There is no need to give examples of this aspect of behaviour of all natural and social systems. Strictly speaking, if one looks merely at microscopic phenomena (and avoiding any philosophical discussion on what 'randomness' really means), one can always find a cause for any particular natural event. However, since these causes are often outside the scope or the spatial or temporal resolution of the problem or if there is no precise data on them, it is often more simple to represent them as a stochastic noise or 'disturbance' of the natural periodicity. This conceptual simplicity corresponds, however, to an increase of the complexity of the mathematical formulation and raises some interesting questions about decision making in non-deterministic conditions, which will be briefly dealt with in Chapter 4.

Complexity

Another characteristic of environmental problems is their complexity in terms of the multiplicity of criteria or points of view under which they can be seen and of the involvement of different social groups. Several different technical means have been suggested to translate these complex facets into the problem formulation, but they have largely failed up to now to produce operative solutions that are acceptable to the decision makers. This is why a more recent and flexible approach (see again Chapter 4) has abandoned, at least in part, the idea of being able to translate an environmental problem into a mathematical optimization program, extending the use of simulation models which can easily answer 'what if' type of questions.

Simulation models are in fact easier to implement, may become very complex without being prohibitive for small computers and allow a direct involvement of the decision maker in the formulation of the hypothetical decisions to test.

Massive data requirement

The last characteristic to be emphasized here is the need for a great deal of data to properly model and verify an environmental problem. This is due to the complexity of the problems already mentioned and to the need for thoughtful experimentation before a model goes into actual use. Environmental decisions are in fact risky decisions, and one obviously wants to reduce the possibility of errors to a minimum.

Data used in these kinds of problems have, in general, a precise structure due to their time and geographical links (see Chapter 3). Time-varying data are usually connected into time-series and thus can be easily represented, for instance as arrays, in any programming language. Static data often describe the physical characteristics of the real system and are usually characterized by their position in space. A record structure, with three fields to describe the space coordinates and a fourth field for the specific value, is a possible data structure and is often used to represent these items. Obviously, there are many other less structured types of data, but they usually comprise only a very small percentage of the two quoted above. For instance, the computerized description of a certain area may be constituted by a regular grid of spatial positions, some time-series of environmental variables associated with specific measurement points, and a set of data representing the region border.

EDSS structure

In adapting the general DSS objectives stated in Chapter 1 to the case of environmental problems, an EDSS can be seen as a computer system that is able to handle models with the characteristics mentioned above; to connect them to create more complex model structures; to assist the user in developing new models following the procedure outlined in section 2.1; to manage the exchange of environmental data; and to include parameter estimation and optimization facilities. Its final aim should be to help the user in finding some structure in the problem to be solved by allowing him or her to screen different well-defined models (the only kind of models the computer can handle) to find those which appear close to the original system and to test on them the effects of some proposed actions. These actions, in turn, may be simply tried by the user or generated by the computer in order to satisfy some user-defined criteria.

To perform these tasks, the architecture proposed in Chapter 1 and presented again in Fig. 2.3 seems to be well suited.

The Database should be designed in order to perform in the best manner when dealing with spatial data and time-series; the Modelbase should

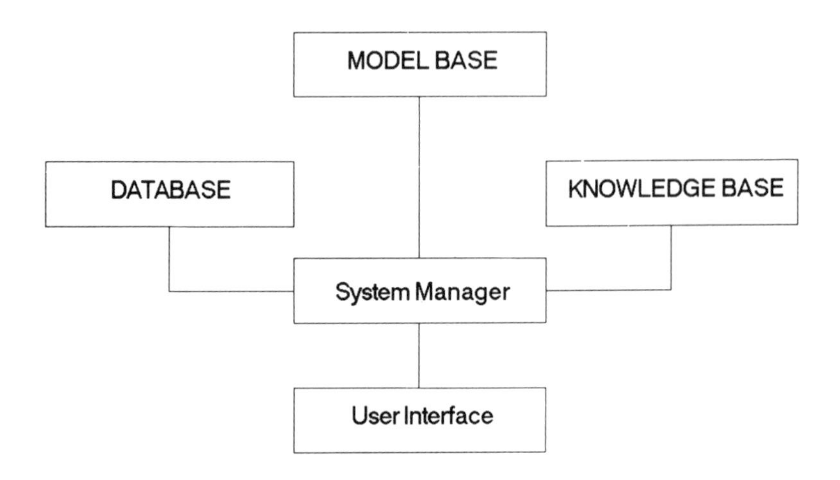

Fig. 2.3 — The architecture of an EDSS.

contain a large set of models of environmental systems; and the Knowledge base should provide all the necessary information for handling models and data, such as how to run each model, its data requirements, and in which cases it can be utilized. Finally, the System Manager should provide the necessary coordination between the various bases in order to create new models, or connect existing ones.

It is important to realize that several different types of expertise must converge to implement such a system: computer experts for defining the bases' structure and the system software; modellers to fit an accurate formulation of different problems in the modelbase; and environmental engineers to guarantee the coherence and the reality of the overall design and to fix a suitable level for the interface. These experts must work in strict cooperation, or the development of such a system will remain, as many others, a 'paper project'.

2.4 USER AND SYSTEM ENVIRONMENTS

In order to discuss the EDSS structure in more detail, it is important to analyse who are the users of the system and its role in an environmental institution.

A computer system of the type proposed above may be the key instrument by which an environmental authority may test the effects of alternative decisions and perform a series of studies on the environmental problems in its area cf interest.

The institution in which this system is to be implemented is supposed to bear full responsibility for a certain set of environmental decisions; it is in a higher position in the hierarchy than all the other parties involved in the real system. That is, there is a unique decision maker (the agency, even if decisions are taken by a group of people within it), and any other entity carrying out any activity in the system has to take the agency's decision for granted. This means also that decisions taken by the agency enter into someone else's problems as external inputs that cannot be modified.

This assumption, which is probably the most realistic one in environmental problems, means that all the multi-decision makers' problems and techniques, for instance 'game' or 'team' theories, will not be dealt with by the proposed EDSS.

Since decisions usually become more complex and difficult in the medium to long range and different aspects of the solution have to be considered, the proposed EDSS is intended basically for planning purposes and not for routine control operations. It is not directly connected to a measurement network or to implementation devices. Occasionally, it may be connected to such existing networks to gather data and information necessary for a specific study, but it does not represent the normal database of the agency. A computer which (at least for software if not for hardware) works offline with respect to the routine activities of the agency can be used. The EDSS is thus a sort of laboratory to investigate environmental behaviour and the effects of man-made actions. In this connection, it may also be the prototyping environment in which other computer applications are tested.

If the problem at hand is a typical planning problem, such as locating a facility, investing in environmental protection actions, or formulating a rule regulating fishing or hunting, the EDSS should provide all the tools needed to devise and to experiment with an acceptable solution which the agency can adopt.

In the case of more complex problems, the final decision may envisage the use of a computer for daily operations. For instance, in order to tackle the problem of reducing the short-distance fall of pollutants due to emissions from a power plant (see Chapter 4 for details on this problem), the computer is used to plan the type of meteorological variables to measure, the frequency of those measurements, the control policy to use (filter emissions, substitute fuel, etc.), and how to implement the policy. EDSS should help in deciding these points, but subsequently the control system must be implemented on a different computer, possibly in a different computer language, with different input–output facilities. The person in charge of the control of the plant will not in fact be the same person who made the decisions about the structure of the entire control system. This is, in a certain way, the approach followed by some recent simulation software packages (for example, Modeller 100, Xanalog, Genesis), which offer a wide variety of facilities to design a simulation or control program but then produce an executable version (in a different computer language) of the model which is much less flexible, even if computationally more efficient.

The user of an EDSS will most probably be an environmental engineer with some detailed knowledge about the problem he or she wants to solve and who must not be involved in software development or hardware problems. The user will be asked to suggest to some political authority or to implement directly, if the social and economic involvement is below certain thresholds, technically sound decisions on a variety of problems in a time span which may vary from days to months. Thus, the user cannot undertake extensive research efforts or data collection campaigns. He or she will often have to operate with insufficient information and with poorly structured problems, owing to their complexity and multifaceted nature. The environmental engineer thus needs an instrument that allows him or her to model the problem and test alternative solutions in a relatively short time span, sometimes postponing the attainment of greater accuracy to subsequent, more detailed studies.

2.5 REMARKS ON HARDWARE AND SOFTWARE

The series of tasks outlined in the previous paragraphs call for some comments on the hardware and software necessary or useful to accomplish them. The computer may be of any type, although a personal workstation connected to the agency's local area network (LAN), as shown in Fig. 2.4, will probably be preferable. Mass storage for data should be largely available, but not comparable with the agency's main environmental data bank. The latter may be on some large device, which need not be extremely fast, since data retrieval (when data are available on a computer medium) is usually just a limited portion of the work needed to solve a problem. Environmental data are suitable for storage on a WORM device, a write-once read-many times removable disk based on laser technology which may offer capacities of several hundred megabytes at a very attractive price. A characteristic of environmental data is that they are static in the sense that new data are always added to a time-series, but stored data are corrected only very rarely, for example when an error is discovered or when an unrecorded value can be replaced by a meaningful value.

A connection to the central data bank is obviously desirable, but this is automatically achieved with a connection to the agency's computer network. The computer will probably use a mouse as the standard input device, but a digitizer may be present to acquire geographical data directly as well as to digitize rapidly old data registered by instruments writing on paper strips.

A direct connection with the telemetering network should be available to test applications based on it, but it will be active only occasionally, since continuous data registration is not a task of the EDSS. It may not be necessary to plan such a connection, because it is quite probable that the telemetering system will also work, possibly through a concentrator, as a node of the overall computer network.

The major output device is represented by a high-resolution colour graphic monitor and by a printer–plotter for hard copies. The possibility of connecting the monitor to a videobeam is useful in order to use the machine as a device to generate and support an open-minded exchange of ideas during a meeting with people involved in the practical problem. A floppy

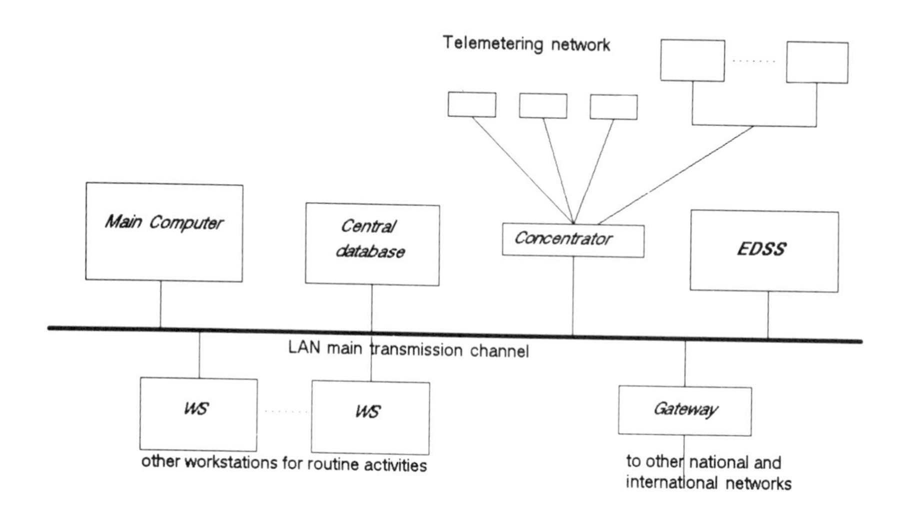

Fig. 2.4 — EDSS connection to an environmental agency's LAN.

disk system will also be required for the input of data from machines outside the network or for transferring the results of the prototyping phase to special-purpose computer systems.

Choosing the kind of software necessary to accomplish this job is usually a twofold problem. On the one hand, one would like to give to the user the greatest possible power and freedom and thus will conclude that the user should simply be given a very powerful computer language with database access and advanced programming facilities (see Chapter 5) (a recent application of such languages for the development of environmental models can be found in Keffer 1988) or, at least, a general-purpose simulation language. On the other hand, one might think that the user should not be involved in programming and program testing but that some power and freedom may be sacrificed to allow him or her to state problems and models in the easiest possible way. In this case, the user will probably always operate within a very high-level software environment in which he or she may find a number of facilities useful in carrying out the job. The interaction will always be through reduced input facilities such as menus, and this will in some way limit the user's power to those functions which have already been prepared in the machine. Obviously, all menus could be modified and enriched, but this can be accomplished only through some programming and thus, in the present case, must be performed by someone other than the user, who has the necessary computer background. Such a software environment will probably also use an artificial intelligence technique to identify the user's needs and to allow a more natural interface. This confirms that any modifications of the software will normally be outside the user's reach.

A feature that would be advantageous in both cases is that of accessing from and embedding into the EDSS software all the models already developed by the agency and available in a computer usable format. The optimum would be to reuse them without the need to recompile and relink them, but this implies a high level of compatibility between the EDSS and the operating systems of the agency's other computers. One possibility (see, for instance, Taylor and Hurrion 1988) is to use the EDSS computer to drive another computer in which the required model is available through the network software and by exchanging messages and files between the two machines. This idea may be particularly attractive if the model the user wants to run is extremely large and thus requires a mainframe in order to get an answer in a reasonable time. However, these cases are normally limited, and the coordinated use of two machines on the network normally slows down the execution speed considerably.

2.6 LEARNING BY EXAMPLE

Given the problems outlined above, it appears impossible to devise a formal procedure for the synthesis of an EDSS. Thus, an alternative approach will be followed here, based on the analysis of several examples which reflect in various ways the general framework presented in this and the previous chapter.

The following four chapters will be devoted to the evaluation of single components of the EDSS (database, modelbase, knowledge base, interface) and will analyse in some detail existing systems in which that component has been particularly well developed. None of those systems fits exactly the definition of EDSS that has been given, and a fairly complete prototype will be illustrated in Chapter 7. However, each of the systems presented has been fully developed in order to fit some real decision context and thus reflects, at least partially, some of the needs of decision makers.

Before closing this chapter, a few examples of actual computer installations within environmental institutions will be given in order to show the range of possible applications and the kind of problems involved. They span the range of problems an EDSS may cover and are important either for historical reasons or because they are representative of new trends in the use of computers in environmental planning and management. Most of these examples are taken from the field of water resources management, since, as already pointed out, the application of computers in this area of environmental science started earlier and spread most rapidly (Rinaldi 1984).

NEDRES: A database of geophysical information

The US National Oceanic and Atmospheric Administration (NOAA) maintains a large database concerning all information on the natural environment, particularly oceanic and atmospheric measurements, in the United States (Barton 1987). The database, called NEDRES (National Environmental Data Referral Service), does not contain the data themselves but rather all the information necessary to check whether they are available and how to get them. The computer storing all this information is located at BRS Information Technologies, Latham, New York, and anyone who has been provided with the necessary NOAA password can access it through a standard telephone link.

The database constitutes a reference for any environmental research and considerably shortens the time which is normally spent searching for information as the preliminary step in any investigation. Furthermore, it combines information from many different sources and thus allows a multidisciplinary approach to any environmental problem.

NEDRES presently contains more than 16 000 data descriptions taken from data catalogues, publications, bibliographies, computer program manuals, users' guides or related documentation. Each data description is quite large and contains a title, an abstract of about half a typewritten page, information about the data collection, the period of interest, the geographical location, a description of the measured quantities, the contact person, the availability and data-related publications. The overall description of a single entry of this type may thus last for several pages. A fairly similar data structure has also been used by Guariso and Werthner (1988) to store information relative to environmental models.

The interesting aspect of NEDRES, from the computer science point of view, is that, being directed to a variety of people with extremely different degrees of computer experience, it has been organized for highly user-friendly interaction. The search for an item can be accomplished by asking for all records containing any English word or set of words. All words contained in an item are indexed (except conjunctions or prepositions such as 'and', 'to' or 'not'). In this way, the text of each description can be in plain English and the retrieval does not need any particular query language. The disadvantage of this approach is obviously in the amount of mass storage needed for the index, which often exceeds the disk occupation of the data themselves. However, the availability of systems like CD-ROM (Compact Disk-Read Only Memory), read-only devices which use standard compact disks as a medium, may bring this kind of application into the range of personal computers.

Instead of publishing printed bulletins, a large environmental agency whose data are of interest to a sufficiently high number of users will probably issue in the future a quarterly or yearly CD-ROM with all environmental data and the required indexes in much the same way as has already been done for some publications of general interest (for instance, the *American Encyclopedia*).

The Cleveland solid waste collection system

At the beginning of the seventies, a severe budget reduction of the Waste Collection and Disposal Division in Cleveland, Ohio, called for a complete revision of the collection system and, in particular, for a change in the crew composition, in the type of trucks and in the general schedule of work (Clark 1973). At that time, the city had a population of over 700 000 and disposed of about 320 000 tons of solid wastes at a total cost of 14.3 million US dollars per year.

The first step towards the reduction of these costs was the establishment of a computerized information system containing all the information about the service from constant data, such as vehicle capacity, cost and expected life, to daily data such as precise time vehicles left and returned to the motor pool, mileage, weight of the collected wastes, etc.

Starting from these data, it was possible to formulate precisely a mathematical model of waste collection in order to reschedule it in a more cost-effective way. The model was coded in the programming language Fortran and used only by computer experts. Owing to limitations of the computer available at that time, the results were reported on paper to decision makers, who had to ask the computer people to check additional scenarios or produce more detailed outputs as required. The computerized planning of the service allowed a cost reduction of more than 6 million dollars per year (Clark 1974). This is often quoted as one of the first examples of the success of modelling and computer techniques in environmental problems (Beltrami 1977).

The River Dee Regulation Scheme

The River Dee has an 1800 square kilometre catchment area in North Wales; during the period 1966–1976, it served for a pioneer research programme on computer-assisted reservoir regulation sponsored by the UK Water Resources Board and by the Welsh Water Authority. It probably has one of the longest records of experience in applying computer and modelling techniques to environmental problems.

The first installation of a computer which connected some telemetering stations dates back to 1972 (Jamieson 1972, Wilkinson 1972, Jamieson and Wilkinson 1972). The purpose of the network was basically to improve knowledge of the current status of the river system in order to regulate the release from four upstream reservoirs (see the map in Fig. 2.5) with a total capacity of 174 million cubic metres. These reservoirs controlled about one-third of the total runoff of the area and were operated for several different purposes. Historically, the first purpose was to supplement the river's natural flow during the dry season in order to allow abstraction of about 10 m^3/s and guarantee a minimum flow at Chester of about 4 m^3/s. More recently, the system has also been used for flood control purposes in the downstream plain and in the town of Chester. Drinking water supply, hydropower production at the reservoir dams, recreation, safeguarding fisheries, and the control of the intrusion of salt water at Chester Weir during high tides are some of the other objectives of management.

The control system is based on a computer centre at Bala, near the major control point of the river. It receives information every half hour from the four reservoirs, 15 flow telemetering stations and five raingauges connected through radio and telephone links. Forecasts of the situation for the next 48 hours (about the time it takes the water to travel from the most upstream reservoir to Chester under low flow conditions) are computed on the basis of simple regressions and water balance models, and, if they are found unsatisfactory in some point of the basin, the calculation is repeated with different hypotheses on the releases from the reservoirs. In this way, the

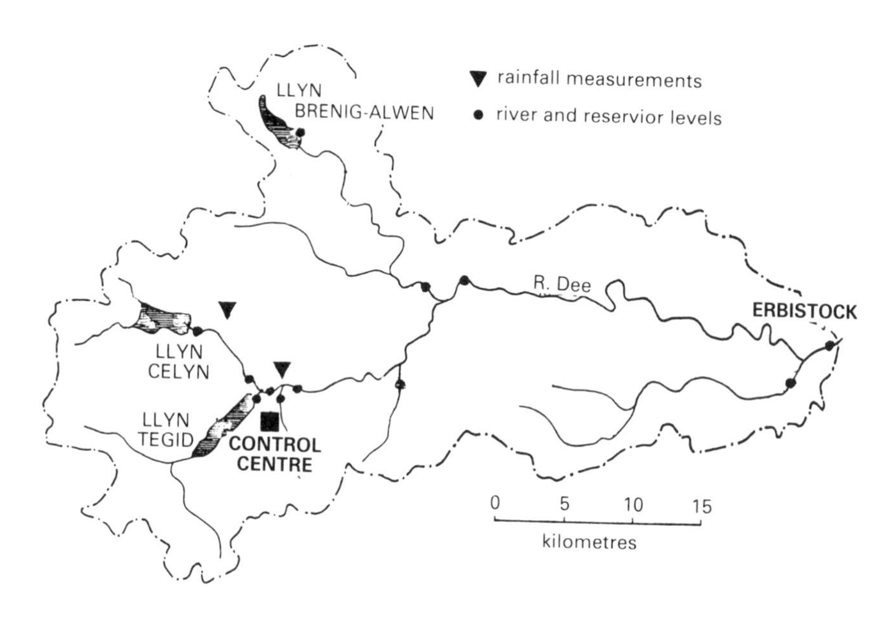

Fig. 2.5 — Map of the River Dee catchment and its telemetering network.

managers can test the effects of alternative solutions before taking their final decision.

The system is under continuous improvement (Lambert and Lowing 1980), and the managers are relying more and more on the information supplied by the computer: in 14 years of operation, the number of floods at Chester dropped from three per year to a single episode!

The Canal de Provence on-line control system

The Canal de Provence is a large artificial canal system in the south of France (see Fig. 2.6). It distributes an average flow of about 40 m^3/s through a canal network of several hundred kilometres, mainly for hydroelectric and agricultural purposes.

The main feature of the system is that the agricultural irrigation network is under pressure and completely 'demand driven', which means that each individual user may freely operate the control structure for a direct supply of water. Thus, the purpose of management is to supply the required water, avoiding losses and minimizing cost, while always satisfying the hydraulic constraints of the system.

Starting in 1970, the system has been automated by mechanizing the main control structures, setting up a telemetering network which reports the situation at many control points to a main computer, and developing models to forecast users' demand. Since the travel time of water within the system is of the order of 6 hours, only variations of the demand within the same day are significant and a 12-hour forecast is sufficient to plan the discharge of upstream reservoirs and for setting the positions of all the control gates. In

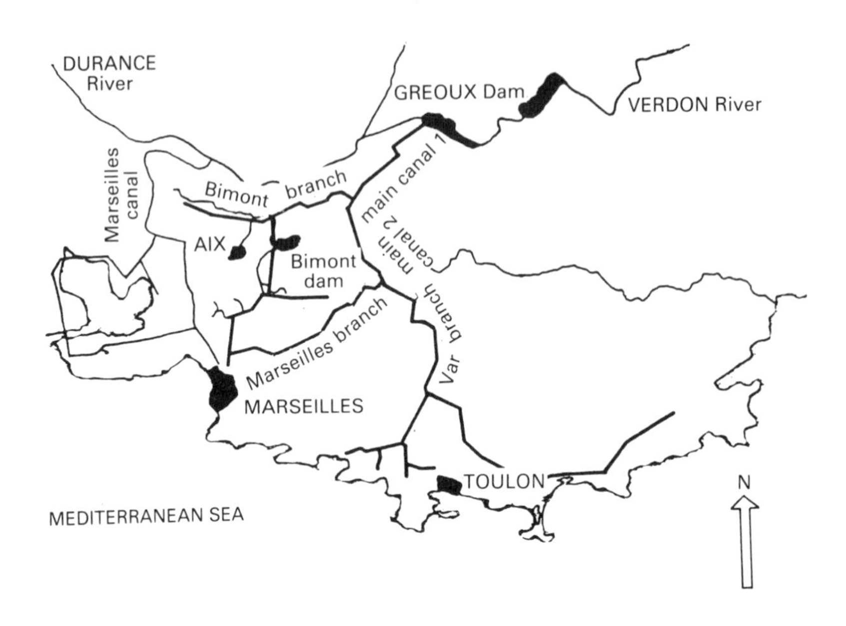

Fig. 2.6 — The Canal de Provence system in Southern France.

this case, the computer acts exactly as a standard process control system, almost without intervention of the operator.

The agency responsible for water supply (Société du Canal de Provence) is quite satisfied with the performance of the forecasting and control system and quotes a maximum departure from the target flows of around 15% (Declaux 1985).

OASIS: An artificial intelligence tool to support water management in South Florida

This application relies heavily on artificial intelligence methods. The system has been designed to support the management of the South Florida Water Management District, which operates more than 200 water control structures along 3200 kilometres of primary canals in a region of about 46 000 square kilometres (see Fig. 2.7). The District operates its storage facilities and canal network in order to prevent flooding during the wet season (June–October) and to supply water mainly for agricultural purposes during the dry season. Other management objectives include environmental and water quality conservation, preventing salt water intrusion, protecting South Florida wetlands, and providing sufficient discharge to a national park. In order to monitor the behaviour of such a complex system, the District is equipped with a sophisticated microwave telemetering network, which processes up to 250 000 records per day from 650 gauges and field personnel observations. This information, together with meteorological forecasts, historical statistics, intuition and 'common sense', provides the basis for daily operation decisions of the District's managers.

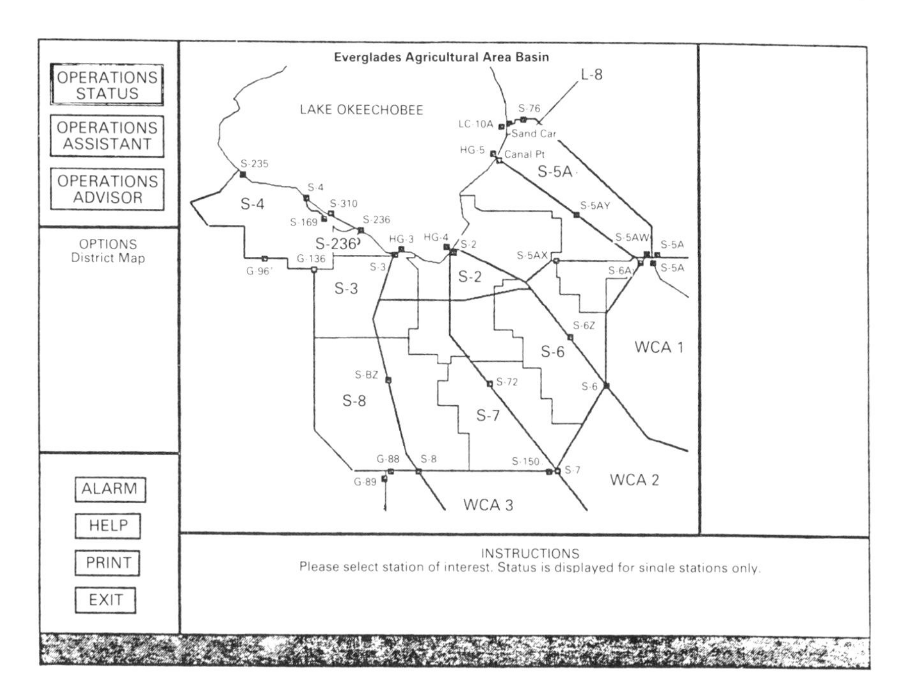

Fig. 2.7 — The South Florida District as shown on the computer screen by OASIS software.

In 1985, the District decided to automate part of the operation procedures, encoding the managers' experience into an expert system called OASIS (Operations Assistant and Simulated Intelligence System) (SFWMG 1987).

The system, which is presently under completion, utilizes special hardware (a Symbolics 3640 computer) and software (ART, Automated Reasoning Tool, an expert system shell from Inference Corp.). This shell translates a rule-oriented user language† into the programming language LISP and offers additional utilities to create a very friendly user interface. The rules used in the system are of the type 'IF a certain situation occurs THEN take a certain decision' and translate the experience acquired by the district managers over a period of 15 years. They represent the control and measurement actions which are undertaken by the agency in all possible situations. These rules are thus very complex and site-specific and submit to the operators a set of decisions for all storage, pumping and canal facilities based on the experience gained on past occasions.

The expert system has been integrated into a more complete information system that provides real-time data display and historical and current data plots, checks actual alarm conditions and forecasts future ones.

† Chapter 5 will discuss these software tools in some detail.

REFERENCES

Barton, G. S. (1987) NEDRES: An interactive computer tool for locating geophysical information. *EOS*, **68**, No. 19, 514–515.

Beltrami, E. J. (1977) *Models for Public Systems Analysis,* Academic Press, New York.

Bundy, A. (1985) Intelligent front ends. In: Bramer, A. M. (ed.), *Research and Development in Expert Systems,* Cambridge University Press, Cambridge, UK, pp. 193–204.

Chow, V. T. (1964) *Handbook of Applied Hydrology.* McGraw-Hill, New York.

Clark, R. M. (1973) Measures of efficiency in solid waste collection. *ASCE J. Environmental Engineering Div.* **99**, 447–459.

Clark, R. M. (1974) Solid waste: management and models. In: Deininger, R. A. (ed.), *Models for Environmental Pollution Control.* Ann Arbor Science Publ., Ann Arbor, Michigan.

Declaux, F. (1985) The role of a forecasting model in the regulation of a water supply system. In: Todini, E. (ed.), *Proc. Int. Symp. on the Role of Forecasting in Water Resources Planning and Management.* University of Bologna, Bologna, Italy, pp. 159–168.

Finzi, G., Lancini, R., Savoldelli, A. & Brusasca, G. (1988). AST-2: A PC package to study causality relations among environmental variables. In: Zannetti, P. (ed.), *Proc. ENVIROSOFT 88.* Computational Mechanics Publications, Southampton, UK, pp. 467–478.

Galassi, S., Gatto, M. & Zanetti, B. (1988) BIOCON: A program for the parameter estimation and the simulation of a simple bioconcentration model. In: Zannetti, P. (ed.), *Proc. ENVIROSOFT 88.* Computational Mechanics Publications, Southampton, UK, pp. 715–724.

Guariso, G. & Werthner, H. (1988) A software base for environmental studies. *Computer Journal,* 31, No. 6, 550–553.

Henize, J. (1984) Critical issues in evaluating socio-economic models. In: Ören, T. I., Zeigler, B. P., & Elzas, M. S. (eds), *Simulation and Model-based Methodologies: An Integrated View.* NATO – ASI Series no. 10, Springer–Verlag, Berlin, pp. 557–590.

Hurst, H. E. (1952) *The Nile: A General Account of the River and the Utilization of its Water.* Constable, London.

IAHS, International Association of Hydrological Sciences (1989) *Proc. of the Third Scientific Assembly.* Balttimore, Maryland. In press.

Jamieson, D. G. (1972) River Dee Research Program 1. Water supply and flood alleviation. *Water Resources Research* **8**, No. 4, 899–903.

Jamieson, D. G. & Wilkinson, J. C. (1972) River Dee Research Program. 3. A short-term control strategy, *Water Resources Research* **8**, No. 4, 911–920.

Keen, P. G. W. (1987) Decision Support Systems: the next decade. *Decision Support Systems* **1**, No. 1, 25–36.

Keffer, T. (1988) Data analysis and interactive modelling system. *EOS* **69**, No. 9, p. 133.

Lambert, A. O. & Lowing, M. J. (1980) Flow forecasting and control on the River Dee. In: *Proc. Hydrological Forecasting Symp.*, IAHS–AISH Publ. no. 129. IAHS, Washington, DC, pp. 525–534.

Maass, A., Hufschmidt, M. M., Dorfman, R., Thomas, H. A., Marglin, S. A. & Fair, G. M. (1962) *Design of Water-Resource Systems*. Harvard University Press, Cambridge, Mass.

Meadows, D. H. & Robinson, J. M. (1985) *The Electronic Oracle*. John Wiley, Chichester, UK.

Morrice, H. A. W. (1958) The use of electronic computing machines to plan the Nile Valley as a whole. *IAHS Publ.* **45**, No. 3, 95–105.

Ören, T. I.(1984) Quality assurance in modelling and simulation. In: Ören, T. I., Zeigler, B. P., & Elzas, M. S. (eds), *Simulation and Model-based Methodologies: An Integrated View*. NATO – ASI Series no. 10, Springer–Verlag, Berlin, pp. 477–518.

Rinaldi, S. (1984) Systems analysis and environmental modelling. In: Ruberti, A. (ed.), *Systems Sciences and Modelling*. UNESCO - Reidel Publ. Co., Dordrecht, The Netherlands.

Sargent, R. G. (1984) Simulation model validation. In: Ören, T. I., Zeigler, B. P., & Elzas, M. S. (eds), *Simulation and Model-based Methodologies: An Integrated View*. NATO – ASI Series no. 10, Springer–Verlag, Berlin, pp. 537–556.

SFWMG (South Florida Water Management District) (1987) *OASIS. The South Florida Water Management District's operations artificial intelligence program*. District Brochure.

Tavares, L. V. & Da Silva, E. (eds) (1986) *Systems Analysis Applied to Water and Related Land Resources*. IFAC – Pergamon Press, Oxford, UK.

Taylor, R. P. & Hurrion R. D.(1988) Support environments for discrete event simulation experimentation. In: *Proc. Europ. Simulation Multiconf., Nice, 1–3 June 1988*. SCS, Ghent, Belgium, pp. 242–248.

Wilkinson, J. C. (1972) River Dee Research Program. 2. A long-term control strategy. *Water Resources Research* **8**, No. 4, 904–910.

Zannetti, P. (ed.) (1986) *ENVIROSOFT 86*. Computational Mechanics Publications, Southampton, UK.

3

Data management

This chapter will present an overview of databases (DB) and database management systems (DBMS). In that sense, it can also be seen as a short introduction to database theory. But its principal scope is to show the reader the advantages of using a DBMS in environmental applications and to give some suggestions for their design and use. The chapter is organized in the following way: after a rather general introduction, several data models will be presented, thus also examining some recent research in this field. A discussion of the role of a DBMS component in an EDDS follows. Geographical Information Systems (GIS) as a specific application of database systems are described in section 3.8. Finally an example of an application is shown.

In the short history of computerized environmental management and planning, data management has played a central part. This does not seem to be astonishing, since computer applications in that field are very data sensitive and intensive. Environmental computer systems have the following characteristics with respect to their data (see also Chapter 2):

- — the data may have distributions in time and space and thus a complicated logical structure,
- — computer storage has to deal with a large amount of data, for example in the case of timeseries about a long range of time,
- — the quality of input data may be influenced by measurement errors,
- — fast data access has to be provided, when the application programs deal with on-line (real time) control.

Previously, input and output data used in environmental applications, as in all other areas, were handled within simple file systems. In such a case, every application program was responsible for its data and their storage. Changes in the structure of stored data often caused the modification of the programs which worked with these data. It might also have been the case that several programs worked on the same data but these were stored separately for each program. In general, maintaining data in such a file system tends to lead to data redundancy and therefore a waste of memory, duplication of work, and possible inconsistency.

As a practical example of the management of information which is needed and has to be maintained in the field of environmental planning, let us consider a regional institution engaged in the control of air quality. This institution will have a large amount of data kept in its computer. These data may represent information dealing with measurement stations, instruments, and measured values of chemical substances, as well as data from the personnel office concerning the employees of that institution. In addition to these data, there is some structural information. For example, some data may also represent relationships between other data, such as quality reports (e.g. the geographical point at which a specific threshold value was exceeded might be stored as a relation of measurement stations and chemical substances) or working crews (which crews are responsible for which stations).

3.1 DATABASE MANAGEMENT SYSTEM

A collection of data as described in the example above (the data with their implied relationships) and stored in a computer is called a database. A more specific definition is given in Teorey and Fry (1982) where a database is defined as a computerized collection of stored operational data that serves the needs of multiple users within an organization or some defined subset of the organization. A database management system is a generalized tool for manipulating such databases. It is made available through special software for the interrogation, maintenance and analysis of data. Its interfaces provide a broad range of languages to aid all users of an organization. It is important to mention that a database is constituted by data which are maintained. Data such as intermediate calculations (for example, a mean of measured values) are purely transient, but are not maintained and therefore not part of the database. An important aspect of a DBMS is that it allows the user or an application program to deal with the data in abstract terms, rather than as physically stored computer quantities (Ullman 1980).

Some advantages of the management of data by a DBMS are presented in the following, by no means exhaustive, list. Not all of these features are automatically attained, but are the product of a well-designed and implemented DBMS:

— Central storage and management give the users a constant view of the data. For example, every application program of a distinct user group might see the same data with the same relationships.
— Consistency of data can be guaranteed. It is assured that different users can attach the database in a consistent state.
— Redundancy of data can be avoided. Data are no longer stored independently by every application program. This reduces the necessary memory capacity.
— Actuality of the data is provided. In the case of the air quality control institution above, all users can access the most recent pollution values.
— Integrity can be checked by the DBMS. If, in the example, a value at some measurement point passes a specific predefined threshold, the system could conclude that a measurement error has occurred.
— Security barriers can be implemented. Not all users will have access to all data.

— Data independence can be guaranteed. A change in the physical storage of the data does not force a change in the application programs.

So-called data independence is reached by passing two levels of abstraction from the specific view of the user (how he or she sees the data) to the physical implementation inside the machine.

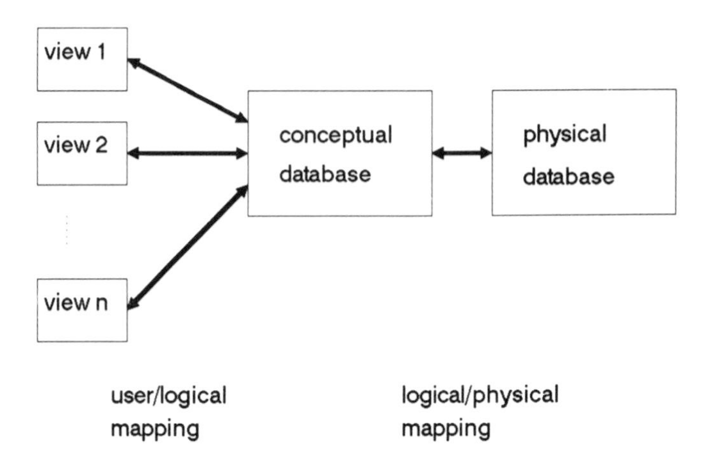

Fig. 3.1 — Levels of data abstraction.

In reality, only the physical database exists. The conceptual database is an abstraction or mapping of the physical storage of the data. The views on their side are abstractions of portions of the conceptual database.

The physical database

The physical database resides permanently on a secondary storage device, such as a disk or tape. The different data structures which are used to store the data on one of these devices will not be discussed here.†

The conceptual database

The conceptual database (it may also be called the conceptual schema) is an abstraction of the real world of interest and, on the other hand, a mapping of the physical database into a logical description. The concepts used to describe the entities of the real world and their interrelationships are defined by a so-called data model.

The classical data models are

— hierarchical models,
— network models, and

† For an extended discussion see, for example, Gotlieb and Tompa 1973, Gotlieb and Gotlieb 1978, Knuth 1973.

— relational models.†

A DBMS offers a data definition language (DDL) to define a conceptual schema, based on the underlying data model. In addition, a data manipulation language (DML) is supported to query and to manipulate (i.e. change) the data. A set of integrity rules, which defines legal states or changes of states of the database (Codd 1980), is also contained in and used by the database management system.

When choosing a data model for a distinct application or an area of application, it is important to note that the structure of a data model should reflect naturally the properties of the real-world objects.

Views

A view or subschema is an abstract model of a portion of the conceptual database. It is defined by using a subschema data definition language which is often similar to or a derivative of the respective DDL. Views may be implemented with respect to different user groups. In the example of the local institution, one group might be the technical department, which has a view that includes all measurement points with their correlated measured values, whereas the personnel office sees only the employees and their working hours with respect to the measurement stations.

Application programs which want to communicate — i.e. find, modify, delete or insert information — with the database management system have to use a data manipulation language or a query language.‡

Some examples of a database query and manipulation may be:

— Find the values of pollution parameters on day dd.mm.yy at the geographical point xy.
— Change the number of hours worked by Mr Who on day dd.mm.yy to the new value a.

An important point is that application programs, once provided with their distinct view, are not forced to modify their access operations when changes in the database's internal structure occur. Data independence is provided by the different logical and physical levels of the database, and application programmers do not have to worry about the maintenance of data.

The following sections give a short overview of the different data models which can constitute a conceptual schema of a database. A simple example will be used describing measurement stations (measuring actual pollution), measurement values (timeseries), geographical features of the controlled area and measurement and/or warning guidelines. A similar structure will be used in the final example in this chapter.

In general, a data model describes a set of entities and their relationships. An entity can be seen as resembling a real-world object. The

† Recently investigated models, such as the semantic data models or the object-oriented models, offer concepts to formalize more semantic information.

‡ Some DBMS differentiate between query language and DML, but it is also possible to integrate the query operations into a DML.

relationships themselves may be stored explicitly as links or as entities. Entities contain attributes or properties, which can take different values. An attribute may represent, for example, the names of the measurement stations having as value the name of a distinct station; another attribute may stand for the geographical coordinates of this station. These entities may also be called records or relations, depending on the specific data model. Choosing a data model thus means choosing:

— a data structure with the correlated entities and their structures,
— operations to manipulate the entities of the database, and
— integrity constraints to correct states of the data.

3.2 THE HIERARCHICAL DATA MODEL

Fig. 3.2 shows an example implementation of the pollution agency problem.

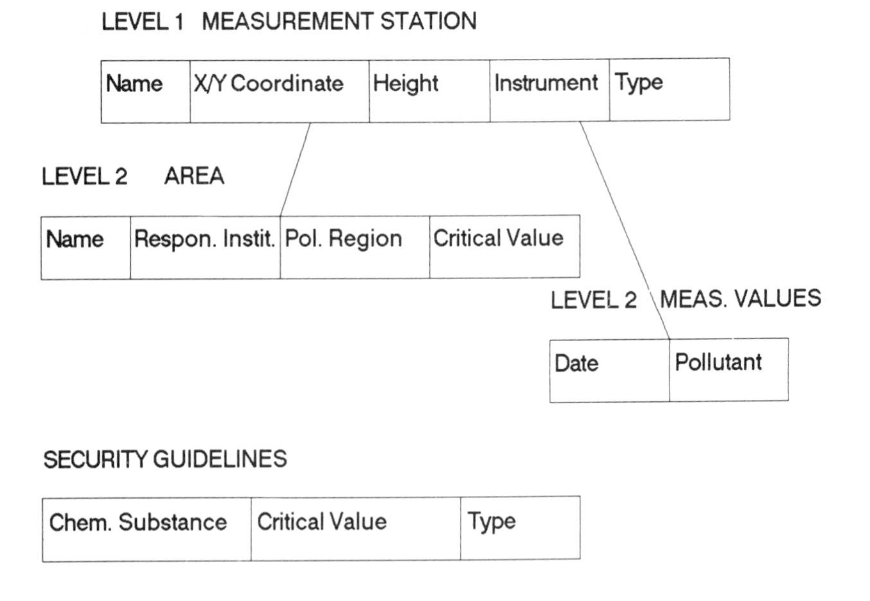

Fig. 3.2 — Hierarchical represenṫration of the example institution. (a) Description of records, their attributes and hierarchy.

Two structures exist: a two-level tree and a single-level structure with only security guidelines. The root of the first structure is constituted by the measurement station; one record is taken for every station. There will be one record for the controlled area, and there will be one area record for every station record, even if an area has more than one measurement station. The second descendent record is the measured value of the sub-

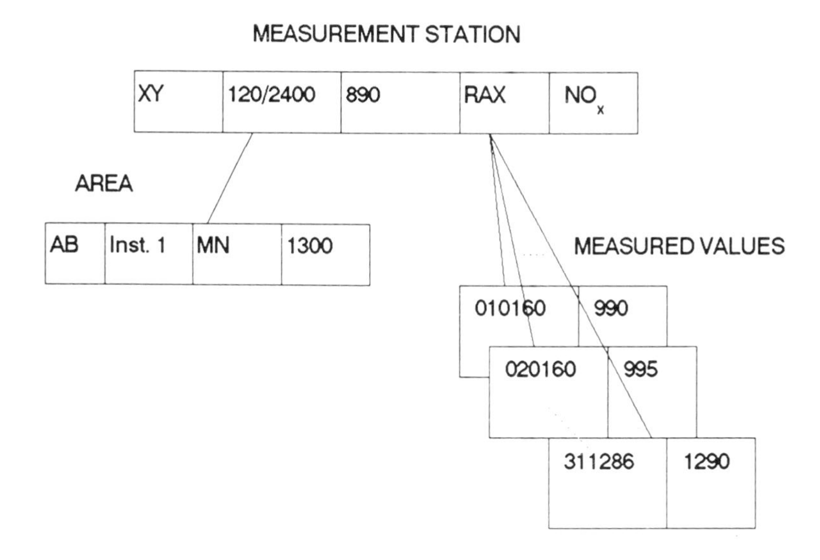

Fig. 3.2 — Hierarchical representation of the example institution.
(b) Instantiated database.

stance, and there will be as many record instances as there are values. In the hierarchical structure, there may be several descendent records for every record, and there may also be more than one instance of each descendent for every record (as in the example of single timeseries records for one control station record, Fig. 3.2(b)). This representation is not the only possible one. For example, the area with several descendent measurement stations could be taken as the root record. The stations, for their part, would have as child record the measured values, thus forming a three-level tree.

Some observations have to be made on the hierarchical data model. First, in the tree representation, some records must exist before others. In the example, the area cannot be represented without having a measurement station: that is, every structure must have a root structure. Second, an area having more than one measurement station must be repeated in the structure for every station. So the area-specific information would be stored several times, with a waste of memory. On the other hand, as will be seen later, hierarchical relationships between single entities do not have to be stored explicitly, since a link drawn between parent and child contains the respective information. In the example the link between station and area reflects the information that every station is situated in an area.

Operations supported by that model are: create an instance of a record, delete an instance, update attribute values within a record, retrieve next instance at the same level (for example, print next measured value of pollutant NO_X), retrieve next child record and retrieve parent record (for example, print measurement station of area AB). The operations in the hierarchical data model are similar to navigation along links. This means that operations on the data structures are powerful as long as one remains inside a hierarchy, but searching outside (in the example going to the security guidelines) and jumping from tree to tree leads to a loss of efficiency.

Constraints are defined on the parent and child records, and normally a record within a level requires a unique key. A key is a privileged attribute which allows a record to be identified. No record can be the child of more than one parent, as a hierarchical data model represents one-to-many relationships.

Some unpleasant features of the data model have to be mentioned. For example, by deleting a measurement station (i.e. root record), the information about an area could be lost if there is only one station per area. On the other hand, updating the area information would mean updating the information everywhere the record appears.

3.3 THE NETWORK DATA MODEL

Fig. 3.3 shows a network representation of the same problem.

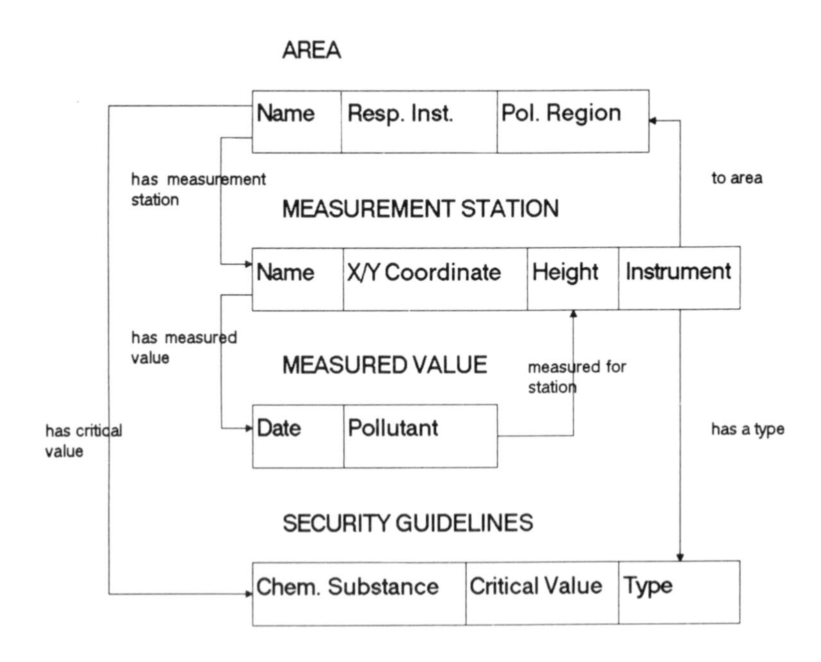

Fig. 3.3 — Network representation.

The network model can be viewed as a generalization of the hierarchical one by allowing a record to participate in more than one relation. As can be seen in Fig. 3.3, a network is a set of entities (records) and named links between them. A main difference from the hierarchical model is that the relationships may be many-to-many, inverse to each other, but only binary, i.e. one link connects exactly two records. So, for example, the links HAS MEASURED VALUES and MEASURED FOR STATION are found between the entities VALUE and STATION. This allows direct access to both the values and the station, whereas, in the hierarchical model, value data could only be accessed via the measurement station. Another difference is that some fields of Fig. 3.2 are replaced by links in Fig. 3.3. For example, the critical value attribute of area is replaced by the link HAS CRITICAL VALUE.

In a network, the many-to-many relationships allow navigation in two ways inside the data structures. This gives the instructions a richer set of possibilities. As in the hierarchical model, the main commands deal with inserting, deleting, modifying and retrieving. The problems related to inserting, deleting or modifying a record are similar to those mentioned in the hierarchical case. For example, changing some field values in an area record would force the user to search for and then to change all of those values in all occurrences of that specific area record.

3.4 THE RELATIONAL MODEL

The relational representation of the example with some instantiated tuples is given in Fig. 3.4.

The relational model has the richest mathematical foundation and the best formulated integrity constraints to define proper states of the database. A formal definition is given by Codd (1970). The structures in a relational model consist only of relations. A relation connects a set of attributes. A relation can be thought of as a table where the columns correspond to attributes and the rows to tuples. These tuples are the entities of interest. Every attribute has its domain, which defines the possible values of that attribute. Every tuple in a database, i.e. every instance of a relation, differs from other tuples. This feature can be compared with the other models, wherein some instances of records have to be stored several times because of structural constraints.

The relational model of the example shown in Fig. 3.4 contains the four relations AREA, MEASUREMENT STATION, MEASURED VALUE and SECURITY GUIDELINES, and every relation has two tuples. It should be noted that the relationship between real-world objects is modelled by attributes in the relations. Thus, there is an attribute AREA NAME in the relation MEASUREMENT STATION to represent the connection between area and station. Also, the attribute MEASUREMENT STATION NAME in the relation MEASURED VALUES reflects the fact that every value belongs to a station.

In the example, the single relations contain more attributes representing the objects of interest. But the relational model has the advantage that every tuple has to be stored only once. Only one tuple would be needed for every distinct area, even if that area had more than one measurement station.

AREA RELATION

Name	Resp. Institution	Pol. Region	Critical Value
Area 1	....	....	....
Area 2	....	....	....

MEASUREMENT STATION RELATION

Name	Area Name	X/Y Coordinate	Height	Instrument	Type
Stat.1	Area 1	....	....	....	....
Stat.2	Area 2	....	....	....	....

MEASURED VALUE RELATION

Measurement station name	Date	Pollutant
Stat.1	....	....
Stat.2	....	....

SECURITY GUIDELINES RELATION

Chemical Substance	Critical Value	Type
Sub.1	....	....
Sub.2	....	....

Fig. 3.4 — Relational representation.

Common operations defined on a relational model are: insertion of a tuple or relation, deleting a tuple or relation, updating of an attribute in a relation, selecting a tuple from a relation, joining two relations based on common values of identical attributes, and selecting a subset of a relation. The well-developed mathematical basis of this data model and these operations show some important characteristics:

— the operations work on entire relations rather than on single tuples, and
— the operations do not depend on the order of the attributes and relations.

Integrity constraints are also well defined in the relational model, and lead to several so-called normal forms for relations. These forms guarantee that certain relationships between attributes of a relation are not destroyed by operations. In the example, the relation MEASUREMENT STATION could be deleted without losing the information about controlled areas, something that could happen in the other models.

3.5 AN OBJECT-ORIENTED APPROACH

The object-oriented approach for modelling real-world entities (or objects) represents an actual development in computer science (see, for example, Stefik and Bobrow 1985, Stefik *et al.* 1986, Banerjee *et al.* 1987, Maier *et al.* 1986). This approach is emerging from activities in knowledge representation in artificial intelligence (Minsky 1981, Bobrow and Winograd 1977), software engineering and simulation languages (SIMULA: Dahl and Nygaard 1966; Smalltalk: Goldberg and Robson 1983). Object-oriented database management systems seem to provide good results in such areas as computer aided design, artificial intelligence and office information systems for handling and representing complex objects such as vehicles, graphical frames, etc. Promising results are also obtained by a combination of knowledge representation techniques and simulation (Doukidis and Paul 1986, Klahr 1986, Guariso *et al.* 1987, 1988), which will have a positive influence on the concept of DSS. One can imagine an object-oriented approach in database systems, in knowledge representation and for a programming language, thus giving the entire system a common conceptual representation, description paradigm and software basis. But up to now only prototypical implementations exist and the area is still the object of intensive research.

As no particular form is currently in use to describe an object-oriented database approach, a few definitions will be introduced at this point. An 'object' is a description of a simple or complex real-world or abstract entity. An object is viewed as an abstract description; it can, in that sense, be seen one level above the data models previewed in the previous sections. One could take as an example a case in which someone wants to model and store information dealing with a police officer and a motorbike: both are objects in a system. The properties of an object are described by variables, which are correlated to them. The real-world object 'motorbike' may have the properties price, speed and front wheel (in addition to others). Thus, the variables of our modelled object are three, one containing a number for the price, one for the speed and one as a reference to another object called a front wheel.

But an object does not have only properties; it may also possess operations that work on these variables. Such operations will be called 'methods'. In the case of the motorbike a method might be the calculation of the actual speed, given the current and previous positions and the time spent in transit. Another concept is of importance in this approach, namely that information between objects is passed via so-called 'messages'. One object, the police officer, might ask for the speed of a passing motorbike via a message, the bike resends the value of its variable, speed, and the policeman could then respond in issuing a ticket.

All objects have a prototypical representation, called 'class'. Structurally equivalent objects correspond to the same class, and they are called 'instances' of this class. Values of variables that are common to all instances of a class (for example, the maximum allowed speed for all motorbikes) are stored inside the class, thus avoiding redundancy. These variables are called 'class variables'; variables which contain instance-specific values are named instance-variables.

Classes participate in a class lattice by subclass and superclass relationships. This type of link is called an 'is-a' link, indicating that class A is-a class B if class A is a subclass of B. A inherits all the variables and methods of B and superclasses of B. This concept provides the possibility of representing the inheritance and orthogonality of objects. The types of links in this approach differ from the links in sections 3.2 and 3.4.†

Fig. 3.5 represents the inheritance lattice of bodies of water correlated

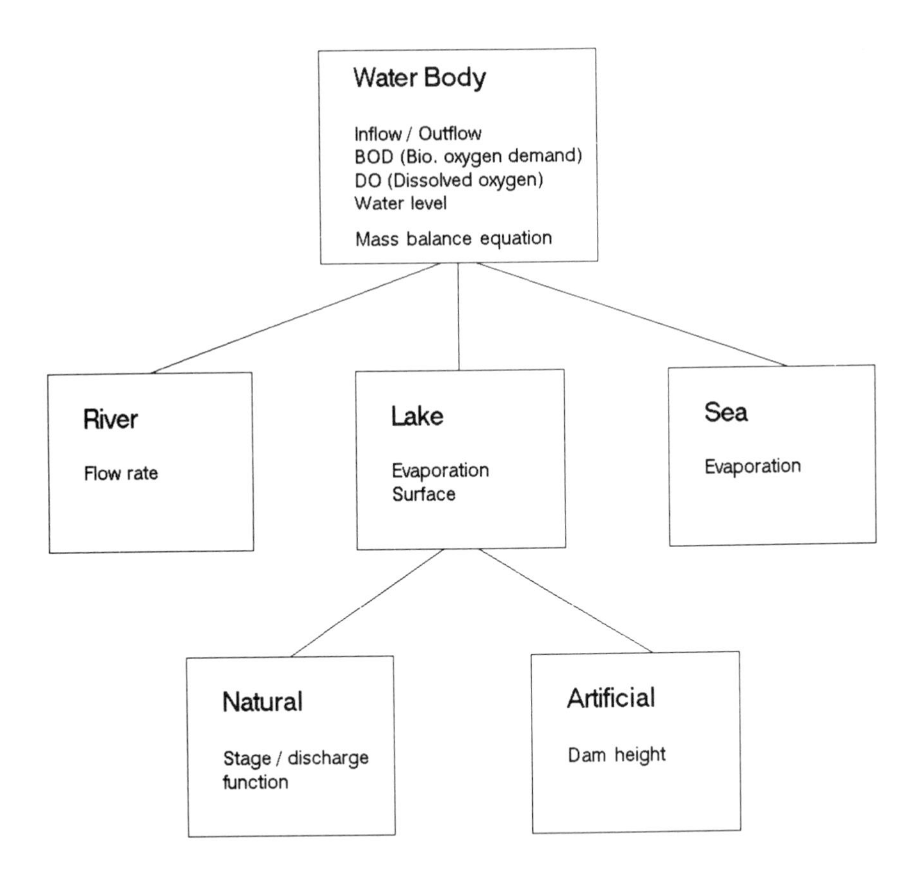

Fig. 3.5 — Inheritance lattice.

with water management down to the level of natural/artificial lakes. In this case, LAKE also possesses BOD values or the mass balance equation from the overall class WATER BODY. Inflow and outflow of the WATER BODY may be references to other objects which may represent timeseries.

† An exhaustive description of this approach will not be given here. The reader will find more on this subject in the literature, listed at the end of the chapter.

The flowrate is a variable which is applicable only to RIVER and cannot be used by the instances of the other classes. It should be noted that discrimination using the orthogonality between LAKE and SEA produces the problem of variable evaporation which is common to both classes. A common superclass, which would be a subclass of WATER BODY, could be introduced to avoid this problem.

Possible operations are the creation and deletion of instances and classes, changing their properties, redefining the sub/superclass structure, retrieval operations using the is-a link relations, etc. The lack of a standard description or definition is, however, a great shortcoming in data manipulation languages, and only prototypical approaches exist. The proposed data structure fits best with an integral object-oriented software engineering approach, using a language such as Smalltalk or LOOPS (Bobrow *et al.* 1985).

There have been attempts to combine an object-oriented data model with a relational one (Fishman *et al.* 1987, Stonebraker 1986) or to interface it to classic high-level languages such as C or Pascal (Purdy *et al.* 1987). This last approach seems to be promising, since not all application programs can be so easily reformulated using an object-oriented language.

3.6 A COMPARISON OF THE DIFFERENT MODELS

The following two criteria (see also Ullman 1980) will be used to evaluate the models described above:

- *Ease of use*. The model must make it easy to formulate queries and to write application programs which want to access the data of the database.
- *Efficiency of implementation*. In large databases the cost of computer storage and search time dominates the effort of implementation. It is necessary to have a model in which the implementation is space-efficient and queries can be handled efficiently.

The criterion of ease of use can best be met by the relational model. The user has to understand only one concept (a relation), whereas the network as well as the hierarchical model require the user to understand records and links. In the hierarchical model complex relationships may mean that an implementation is not straightforward.

In addition, the well-defined mathematical basis of the relational data model has contributed to the development of a set of high-level languages for operating on databases. The other models do not possess languages of that level.

With regard to the efficiency of implementation, the network and hierarchical models may show some advantages. In the case of measurement stations, an example that will be found in a modified form in the final section of this chapter (section 3.9), describing an existing database, the hierarchical model would be the most efficient, considering daily data for 20 or more years and several controlled chemical substances. In the hierarchical model, the name of the measurement station is stored only once on the parent record, whereas in the relational model it would be necessary for every daily value (see Fig. 3.6). In that case, the circumstance that data very often contain a hierarchical relationship can be used.

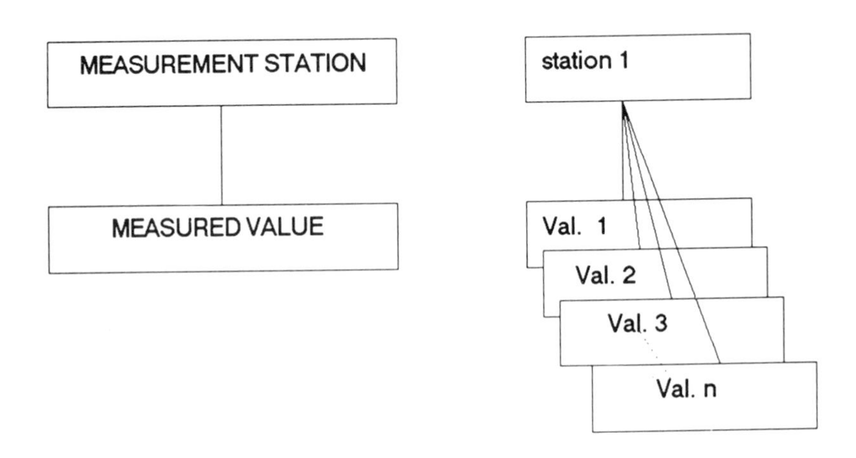

Fig. 3.6 — Comparison of hierarchical and relational models.
(a) Hierarchy and instantiated values.

Name of station	Value
station 1	Val. 1
station 2	Val. 3
⋮	⋮
station n	Val. n

Fig. 3.6 — Comparison of hierarchical and relational models.
(b) Relational representation.

An evaluation of the object-oriented data model, similar to the other models, does not seem to be possible, since up to now no real system based on that approach is available. It can be seen that the approach fits well with systems based entirely on an object-oriented paradigm and will give the user a fairly natural interface to handle real-world objects. In the case of an EDSS, all components, as described in the architecture in Chapters 1 and 2, could take advantage of an object-oriented technique. This is due to the fact that this approach is more a general philosophy of how to see and how to implement real-world circumstances. Thus, all areas of computer science, such as database design, software engineering or artificial intelligence, can be based on this paradigm. Since an EDSS is seen as a set of several components, technically derived from different fields of computer science, it could take advantage of such a common basis.

3.7 THE ROLE OF A DBMS IN AN EDSS

Two types of database applications can be distinguished in the field of decision support for environmental problems.

The first type can be best described as a stand-alone solution. This is probably the most commonly used application today. In this case, the central storage and maintenance is sufficient for the needs of a decision maker. In addition the data manipulation language may possess quite expressive power and thus allows an invocation of simple procedures on the data, as for example some simple statistics. Examples of this type are Geographical Information Systems, discussed in section 3.8, or the database in section 3.9. In the latter section, which describes a project in the Valtellina region in Northern Italy, it will be seen that, for simple managerial decision making, it is sufficient to know the timeseries of some measured precipitation values with some calculated statistics on these series. In this case there would be no need for an integrated complex model base, which would rather complicate the decision finding system.

The second type of application is that introduced in Chapter 1. From an architectural point of view, the database is one component in an integral system. It not only serves as a central and well-organized memory device but also may be useful as an interface or common blackboard for the other parts of the system. The following important features and tasks of a database in an EDSS can be distinguished:

— It supports the central maintenance and integral management of data. The user is supported by a unique interface to the whole system and can invoke database queries during a session with the EDSS without leaving the system. Model calls could be made, for example, depending on the result of a query. The user learns to handle one system with one interface and can remain inside the system until his or her session is finished.
— The database serves as an interface to external data sources. Data extraction from these sources is performed by the DBMS. Supplementary simple integrity checks can be performed after such extractions. In this way, these external sources can be integrated into the system and the user does not have to take care of the interfaces to the system's external world.
— Inside the EDSS, the database management system solves the task of data reduction and the abstraction of large amounts of data. This involves the combination, aggregation and subsetting of entities (by this creating new entities and relationships) in the database, which might be used not only directly by a user but also by a program of the model base.

— The different models in the modelbase have a consistent and an actual view of the data. They are provided with an abstract representation of real-world objects rather than with physically computer-stored quantities. This also allows for an incremental growth of the model base, where sometimes models on a higher level of complexity may integrate already existent models. These interrelated models are not concerned with their interfaces to the data, since this may be managed by the DBMS.
— The database system can handle different types of data, i.e. timeseries or simple scalars. On the other hand, a distinction between measured and computed data can be drawn, thus giving some confidential measures to the different data.
— The database component might also serve as an interface between the different components of the EDSS as well as distinct models in the modelbase. Fig. 3.7 exemplifies this interface task.

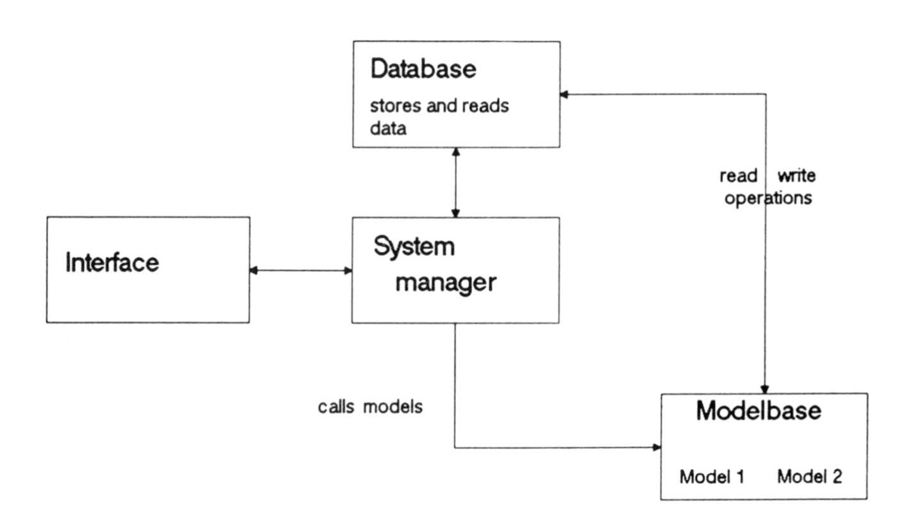

Fig. 3.7 — Database as an interface between models in an EDSS.

In the architecture shown in Fig. 3.7, the user has access to the system via the interface and the system manager or control unit.† This component may store user-supplied data in the DB and invoke one or more models in the modelbase. They communicate via the DB component by writing and reading data. Finally, the user can access the final results. It is worth mentioning that other DBMS features, which have not been described up to now, also have to be used in this rather complex architecture — for example, recovery procedures in the case of a system breakdown or the synchronization of data access by multiple jobs.

† The knowledge base is not represented in the figure; for its specific tasks, see Chapter 6.

3.8 GEOGRAPHICAL INFORMATION SYSTEMS

Geographical Information Systems (GIS) play a special role in the context of environmental database systems. With respect to the classification used in the preceding section, they fall into the category of stand-alone systems, even if their data analysis functions show expressive power. But commands are limited to operations on the database data only, and they do not contain a modelbase for supplying further functions.

Historically, automated or computer-aided spatial data handling was used in the field of cartography with emphasis on display rather than on analysis. Today, owing to satellite remote sensing facilities and increased computing technology, it is possible to obtain topographic, administrative, socio-economic and environmental data for considerable areas of the world. As early as 1977 there were hundreds of systems for handling spatial data (Chock *et al.* 1981). This large number of systems has led to a wide range of variations in geographical data management and handling, and up to now, sufficient standardization is still missing.

Two tasks have to be fulfilled in geographical databases: first, the description of the spatial distribution of an object; and secondly the storage of the thematic information related to the object. A geographical object might be, for example, a political district and the thematic information may deal with agricultural or industrial uses. The two-dimensional digital data structure can be classified into two major classes - grid and topological structures.

In a grid-type structure, an area is subdivided into a rectangular grid. Every grid cell or pixel of an image is connected to a description containing the respective thematic information. This information may cover more than one theme and is thus subdivided into fields, which may take an average or representative value for the respective cell. The size of a cell depends on the implementation and on the spatial variation of the information.

The topological structure, on the other hand, is by far the most widely used. Basic data structures are points, lines and polygons. Attributes are associated with them as shown in Fig. 3.8.

Four main functions of a GIS can be distinguished (see also Tomlinson *et al.* 1976):

- *Data entry*: GIS normally support different hardware devices, such as digitizers or scanners. Interfaces to remote sensing devices may also exist. Additional map editing and updating features are provided. Data held in grid form may be converted to topological format and vice versa.
- *Data management*: This part of a GIS handles the data structures in a grid or topological format, and a linkage is drawn to the thematic attributes of the geographical data. Supplementary information, which lies hidden in the data structure, is automatically updated. For example, in the case of a topological structure, for every polygon pointers are inserted to the left and right side areas.
- *Data analysis*: In addition to traditional data retrieval on correlated

Polygon Topology:

POLY. No.	ARC No.	Label – ID
1	1, 2, 6	– –
2	1, 3, 4	101
3	2, 5, 3	102
4	5, 6, 4	103

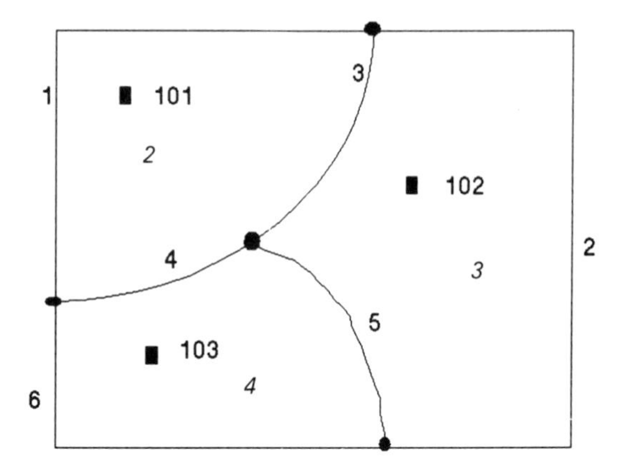

Polygon Attribute File:

AREA	PERIMETER	COVER No.	COVER – ID
9.28	6.30	1	– –
3.1	3.68	2	101
4.05	4.75	3	102
2.13	3.23	4	103

Additional INFO File:

COVER – ID	VEG	SOIL	SLOPE
101	A	10	11
102	B	20	16
103	A	8	17

Fig. 3.8 — Topological data with correlated attributes from Green *et al.* (1985) describing the GIS ARC/INFO.†

†ARC/INFO is a registered trademark of ESRI, California.

thematic attributes, supplementary operations are defined for geographical data. These may include

- overlay of areas
- dissolving of areas
- calculation of distances
- calculation of areas
- interpolation

- transformation of coordinates
- calculation of visibility
- statistical calculations

A typical retrieval, including an overlay operation, may look like this: 'Show the area with agricultural usage where the pollution is beyond some threshold value'.

— *Data output*: Graphic output is normally produced relating cartographic data (i.e. the topological components) to the thematic attribute data which are held in the database. Hardware devices include graphic printers, plotters and high-resolution screens.

3.9 THE ALTA VALTELLINA PROJECT

The Valtellina project is an example of the use of a database in the field of environmental management. In this project, the DBMS represents a stand-alone application, even though the implemented statistical procedure calls can be seen as very simple model invocations. But the main scope is central data storage, maintenance and retrieval functions. By fulfilling these classical DBMS functions, most of the needs of the institution could be met.

Alta Valtellina is the catchment of the upper stretch of the river Adda before it flows into Lake Como (Fig. 3.9); due to its great natural beauty, it is partly included in the oldest Italian National Park, which in the last few years has undergone considerable tourist development. Its water resources have been used, mostly for hydropower production, since 1910. The hydraulic system is managed by the Energy Agency of the Municipality of Milan (AEM) and includes seven hydropower plants with a total power of 840 MW, more than 100 km of canals, two major reservoirs with a total capacity of 187 million cubic metres and several small reservoirs, distributed over a territory of about 800 square kilometres.

Water quality is a major problem, especially during low-flow and holiday periods; the recreational use of the river is also important, due to the increase in tourism in the region. A research project at the Politecnico di Milano focused on Alta Valtellina water resources. The project included the development and calibration of models for water quality simulation and multi-objective water resources planning.

Of central importance for the success of this project and the related water-management tasks of the energy company was a database system (Gandolfi and Werthner 1987). On the one hand it provided a unique data source for all of the researchers involved in the project, and on the other hand it could be further maintained by the company and used by other local authorities for planning and management purposes. The database is implemented on a Personal Computer and it contains timeseries data for a 20 year period, i.e. data registered daily in 17 hydrometric stations (flowrates),

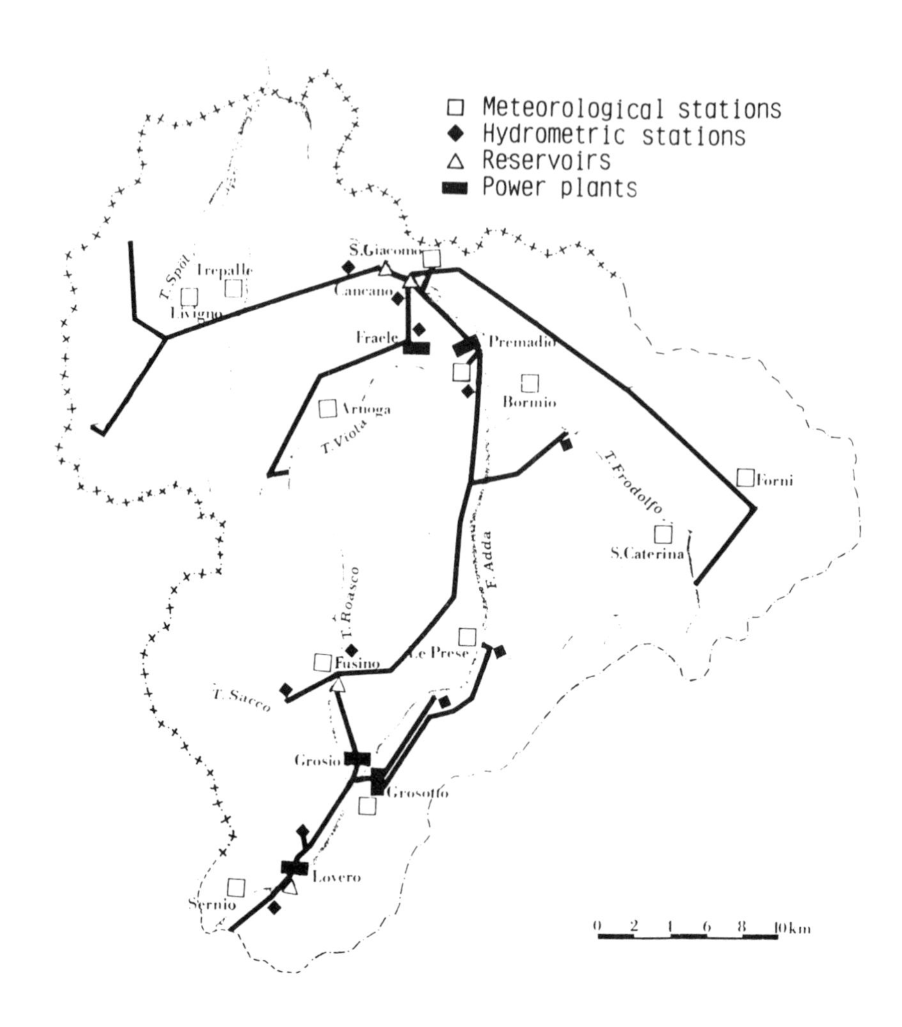

Fig. 3.9 — Alta Valtellina catchment.

three reservoirs (volumes) and 12 meteorological stations (precipitation, maximum and minimum temperature, depth of snow cover) for a total of about 350 000 data.

Analysis of requirements

The database system manages both timeseries data, such as hydrological and meteorological information, and scalars, such as technical and topographical data on river and canal networks and hydrological, hydraulic and geomorphological data on the watersheds. Timeseries data can have different time-steps; a quality flag indicates measured and computed data and missing data handling is supported. A set of procedures performs data maintenance and retrieval. They can be divided into transactions (input, modification and deletion), which modify the state of the database, and reports, which do not imply modification of the state of the database.

From a procedural point of view, there is a great difference between timeseries data and stable data (data which generally do not change in time): transactions concern mainly timeseries data, since stable data seldom need to be modified; reports are of interest both for timeseries and for stable data. A list of the main procedures is presented in Table 3.1.

Table 3.1 — List of the main procedures

Reports
R1 : retrieve timeseries data for one or more given measurement stations;
R2 : retrieve information on measurment station (e.g. location, elevation, shape, size, max. flowrate of a hydrometric station);
R3 : retrieve timeseries data for the measurement stations related to a given station (e.g. flowrates in the canals feeding a given reservoir);
R4 : list the measurement points related to a given one (e.g. the meteorological stations located a given catchment).
Transactions
T1 : maintenance (input, deletion, modification) of timeseries data;
T2 : maintenance (input, deletion, modification) of stable data.

Conceptual design

Two basic entities can be distinguished in the conceptual design, one that describes a measurement unit and one that represents the different timeseries at that point. Since these two entities showed a hierarchical relationship, a hierarchical data model was chosen. The two entities were connected by a one-to-many relationship, and thus the name of the station did not have to be stored for every measured value, as would have been necessary in the relational model. The model chosen reflects the existing structure in the data quite naturally. An example of the two types of entities can be found in Table 3.2.

Additional relationships were introduced which were derived from the processing requirements and define logical links among entities. For example, a relationship exists between canals and reservoirs: certain canals feed

ENTITY: **Meteo – Stat**	Meteor. station
Fields:	NAME HEIGHT (elevation about sea) LATITUDE LONGITUDE ME – PAR (measured param.) INS_TY (techn. characteristics of instruments) START (date of data storage)

ENTITY: **Meteo – Time**	Meteor. timeseries
Fields:	DATE PRECIPITATION T – MAX (maximum temp.) T – MIN (minimum temp.) SNOW (depth of snow – cover)

Table 3.2 — Sample description of two entities.

certain reservoirs. Fig. 3.10 is a simplified sketch of the conceptual schema of the database in which only entities (rectangles) and relationships (solid lines) are shown.

Implementation

For practical reasons, a commercially available DBMS (FOCUS†) was chosen. This product is available on Personal Computers and mainframes, enabling future integration in a larger and more complex system. Furthermore, FOCUS has a powerful statistics and graphics component, which was particularly useful. It supports a hierarchical data model, which meets the requirements of the project quite well. The data are organized in separate structures, one for each kind of measurement station plus one containing information about the whole catchment. Timeseries, which describe the single measurements at a specified geographical point, are correlated with these entities (see also Fig. 3.10 and Table 3.2). Supplementary information is kept at the level at which physical contact is possible between the single entities; for example, a canal keeps the information on which reservoirs are fed by it.

The interface is written in the DBMS data-manipulation language, and consideration was given to ensuring that the final user is completely free from programming burdens or formal query language constraints. The program is menu-driven and, where specific information is needed from the user, he or she uses fill-in forms.

† FOCUS is a registered trademark of Information Builders, New York.

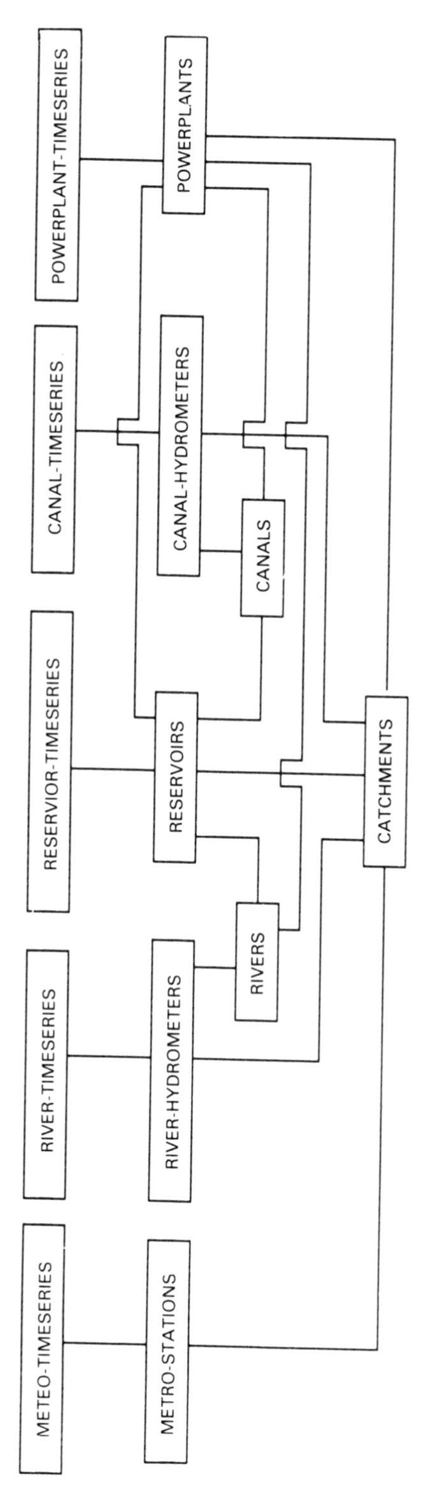

Fig. 3.10 — Conceptual design.

A short overview of the retrieval and analysis function of the DBMS is given below. In fact, with fewer constraints the data analysis module could be seen as a small modelbase. But as there is no possibility for automated sequencing of the single statistical procedures, for example, this DBMS implementation is not considered as a decision support system.

Reports, using different graphical output forms, can be created concerning:

— the drainage network (e.g. which rivers feed a given reservoir; the hydraulic characteristics of a canal);
— the hydrological network (e.g. which meteorological stations are located in a specified catchment; information about measurement stations such as geographical coordinates, elevation, type of instruments, measured quantities, etc.);
— measurement stations and related timeseries (e.g. the values of one or more timeseries variables for one or more given measurement stations).

Different output devices and report formats are available. Monthly and yearly values may be calculated, and the original and calculated values can be shown as connected point plots (Fig. 3.11).

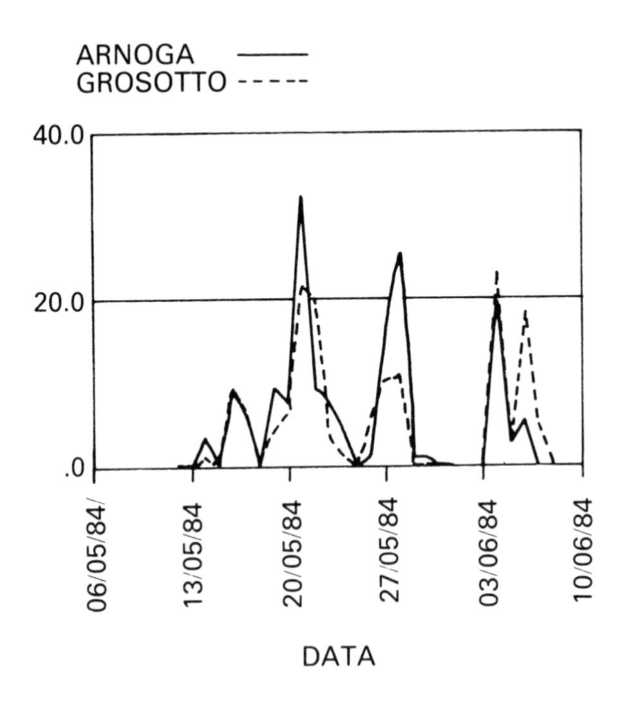

Fig. 3.11 — A daily precipitation plot for two meteorological stations (i.e. Arnoga and Grosotto).

The data analysis module consists of a number of procedures to perform statistics on the timeseries data extracted from the database. Since extraction is the most time-consuming operation, a directory of extracted data was created for each user of the system. The data local to a user are automatically maintained by special procedures, thus leading to a two-level database with a remarkable increase of efficiency. Statistics included in the system are:

— regression analysis,
— analysis of variance,
— discriminant analysis,
— factor analysis,
— exponential smoothing and forecasting, and
— miscellaneous operations on timeseries, including moving-average, interpolation and curve fitting.

The system was very useful during the research project, not only in providing common storage of data, but also in defining a common structure and notation of the real-world objects handled. Moreover, it is currently in use at the Energy Agency.

REFERENCES

Banerjee, J., Chou, H., Garza, J. F., Kim, W., Woelk, D. & Ballou N. (1987) Data model issues for object-oriented applications. *ACM Trans. Off. Inf. System* **5**, No. 1, 3–26.

Bobrow, D. G. & Winograd, T. (1977) An overview of KRL, a Knowledge Representation Language. *Cognitive Science* **1**, No. 1, 3–46.

Bobrow, D. G., Kahn, K., Kiczales, G., Masinter, L., Stefik, M. & Zdybel, F. (1985) CommonLoops: Merging Lisp and object oriented programming. *Intelligent Systems Laboratory Series ISL-85-8*, Xerox Parc, Palo Alto, California.

Chock, M., Cardenas, A. F. & Klinger, A. (1981) Manipulating data structures in pictorial information systems. *Computer* **14**, No. 11., 43–51.

Codd, E. F. (1970) A relational model for large shared data banks. *Comm. ACM* **13**, No. 6, 377–387.

Codd, E. F. (1980) Data models in data base management. *Proc. Workshop on Data Abstraction Data Bases and Conceptual Modeling*. ACM, 112–114.

Dahl, O .J. & Nygaard, K. (1966) SIMULA — An Algol-based simulation language. *Comm. ACM* **9**, 671–678.

Doukidis, G. I. & Paul, R. J. (1986) Experiences in automating the formulation of discrete event simulation. *Simulation Series* **18**, No. 1, 79–90.

Fishman, D. H., Beech, D., Cate, H. P., Chow, E. C., Connors, T., Davis, J. W., Derrett, N., Hoch, C. G., Kent, W., Lyngbaek, P., Mahbod, B., Neimat, M. A., Ryan, T. A. & Shan, M. C. (1987) Iris: An object-oriented Database Management System. *ACM Trans. Off. Inf. Systems* **5**, No. 1, 48–69.

Gandolfi, C. & Werthner, H. (1987) A hydrological information system for the Valtellina region. *Environmental Software* **2**, No.2, 89–99.

Goldberg, A. & Robson, D. (1983) *Smalltalk-80: The language and its implementation*. Addison-Wesley, Reading, Massachusetts.

Gotlieb, C. C. & Gotlieb, L. R. (1978) *Data Types and Structures*. Prentice-Hall, Englewood Cliffs, New Jersey.

Gotlieb, C. C. & Tompa, F. W. (1973) Choosing a storage scheme. *Acta Informatica* **3**, 297–319.

Green, N. P., Finch, S. & Wiggins, J. (1985) The 'state of the art' in Geographical Information Systems. *Area* **17**, No. 4, 295–301.

Guariso, G., Hitz, M. & Werthner, H. (1987) An intelligent Simulation Model Generator. *Technical Report, Politecnico di Milano 87 - 053*, 15 pp.

Guariso, G., Hitz, M. & Werthner, H. (1988) A knowledge based simulation environment for fast prototyping, In: Huntsinger, R. C., Karplus, W. J., Kerckhoffs, E. J. & Vansteenkiste G. C. (eds), Simulation Environments. *Proc. European Simulation Multiconference 1988, Nice, France, June 1988*, SCS, pp. 187–192.

Klahr, P. (1986) Expressibility in ROSS: An object-oriented simulation system. *Simulation Series* **18**, No. 1, 136–139.

Knuth, D. E. (1973) *The Art of Computer Programming, Vol. 3, Sorting and Searching*. Addison Wesley, Reading, Massachusetts.

Maier, D., Stein, J., Otis, A. & Purdy, A. (1986) Development of an object-oriented DBMS. *OOPSLA Proceedings 86*, ACM, pp. 472–482.

Minsky, M. (1981) A framework for representing knowledge. In: Haugeland, J. (ed.), *Mind Design*. MIT Press, Cambridge, Massachusetts, pp.95–128.

Purdy, A., Schuchardt, B. & Maier, D. (1987) Integrating an object server with other worlds. *ACM Trans. Off. Inf. Systems* **5**, No. 1, 27–47.

Stefik, M. J. & Bobrow, D. G. (1985) Object-oriented programming: Themes and variations. *AI Mag.*, Winter, 40–62.

Stefik, M. J., Bobrow, D. G. & Kahn, K. M. (1986) Integrating access-oriented programming into a multiparadigm environment. *IEEE Software*, January, 10–18.

Stonebraker, M. (1986) Object management in POSTGRES using procedures. In: Dittrich, K. & Dayal, U.(eds), *International Workshop on Object-Oriented Database Systems*, IEEE Computer Press, pp. 66–72.

Teorey, T. J. & Fry, J. P. (1982) *Design of Database Structures*. Prentice-Hall, Englewood Cliffs, New Jersey.

Tomlinson, R. F., Calkins, H. W. & Marble, D. F. (1976) Computer handling of geographical data. *Natural Resources Research Report No. 13*. UNESCO Press, Paris.

Ullman, J. D. (1980) *Principles of Database Systems*. Pitman, London.

4

Models for supporting environmental decisions

The role of the modelbase is of central importance in the development of an EDSS. Models represent the information processor, which must extract all the details of the problem under study and make them explicit. Although it is common to see environmental databases working as self-standing systems, it is much more rare to find modelbases in operation. Even if it is not connected to a data and a knowledge base, a suitable modelbase is an important support to any environmental decision.

Most environmental scientists and practitioners agree about the importance of models in solving environmental problems; this can be seen in the flourishing scientific literature in the field and the steady increase in the number of journals in this area.

In spite of the massive research efforts, the application of environmental models, particularly in governmental agencies, has not been so extensive. One of the basic reasons for this is the lack of simple and effective forms to retrieve pertinent information. In particular, there is no easy way to screen existing models to understand which is suitable for a specific purpose, what data it requires, how parameters values must be set, etc. This is exactly the purpose of a modelbase: it should enable the user to access and run a stored model, without the burden of hardware and software problems and, as far as possible, of data preparation.

Thus, the EDSS system manager must coordinate the model and the database in such a way that direct reference to data is embedded into the model definition. Once this is accomplished, the user only has to enter, for example, the name of the system and the initial time of a certain simulation and the system software can choose all the required data starting from that date from the database.

The modelbase must also be connected to the knowledge base, where general information on each model, such as data requirements, data format, software specifications and in general instructions on how models must be used, may be stored. Furthermore, some knowledge about the peculiar features of each model may assist the user in choosing the model that is most suited to a specific problem and help him or her in setting the right parameters or the proper connections with other models. It is possible, in fact, to formalize this knowledge in the form of rules (see Chapter 5) so that new users can benefit from the experience accumulated during previous applications of these models.

The purpose of the present chapter is to consider some possible structures for a modelbase, to analyse the major characteristics of the models normally used to solve environmental problems, and to propose a common framework to integrate all the, apparently very different, types of models. The final section is a presentation of four software systems that have been used in practice for supporting environmental decisions in very specific domains.

4.1 THE STRUCTURE OF THE MODELBASE

The literature provides few studies on possible structures of a model base (Zeigler 1984) and few examples of computer systems for storing and retrieving information about mathematical models that would facilitate their access and use. In some cases modelbases have been developed for a very wide target audience with quite different, and usually rather basic, model and computer backgrounds. Emphasis was more on the retrieval of information on model availability and on transfer from the original source than on program implementation and execution. These modelbases concentrated on structuring information that would help the decision maker in screening a certain number of models and then referred the reader to more complete sources of information on use. When the audience is more restricted and homogeneous and the number of models in use is not very large, less attention needs to be paid to the screening process and the modelbase can be completely integrated into a unique software package.

There is obviously a large range of possibilities, both to assist the model search process and to allow an interactive execution of the selected items within the same software environment. A suitable compromise between these two needs depends upon the amount of responsibility the environmental agency has (and thus how many models are currently in use) and the number of people involved. Chapter 7 will show one of these possible intermediate architectures.

4.1.1 Large modelbases for different users

A report by the National Technical Information Service, US Department of Commerce (1977) is one of the first examples of a compendium of software for environmental studies, with entries on 418 computer programs, from simple hydraulic calculations to models of complex water resources systems. It is not available on a computer, however.

The IAHR (International Association for Hydraulic Research) (1978) computerized a catalogue of 160 programs from 28 different institutions using a technique similar to a database, in which each program was described in about 10 lines of text. The format of the description has been standardized to the following entries:

— Model name

— Keywords
— Purpose of the model
— Method used
— Programming language
— Program type (main, subroutine, package, etc.)
— Computer requirements (type, core, etc.)
— Documentation (existing or not, poor, average, detailed)
— Developing institution
— Charge for use

These items are clearly insufficient to give even a rough idea about the possibility of using a model for a particular purpose; the basic information that can be obtained from the catalogue is the existence of a model of a certain class.

A similar catalogue was prepared by the International Commission of Irrigation and Drainage (Kohlaas 1982) and refers to 187 computer programs available in various institutions of a quite different degree of complexity, indexed under their respective topic (hydrology, economy, groundwater, etc.).

Better inventories have been compiled by the University of Michigan (Richardson *et al.* 1980) on motor vehicles and transportation systems and by the National Technical Information Service, US Department of Commerce (Shriner *et al.* 1978), on all the software packages available in the laboratories of the Department of Energy. The latter is the most extensive inventory; it is fully computerized and contains 820 entries, 255 of which are models, while the others are mainly graphic software packages and data sets. The entries for each model contain the following information:

— Contact person for additional information
— Spatial coverage (field, local, regional, etc.)
— Time span (short, medium, long term)
— Status (presently in use or not, and where)
— Media for transfer (tape, floppy diskettes, etc.)
— Available data sources for input
— Abstract (15–20 lines of text)
— Keywords

They also include the name of the model, the institution that developed the model, method, computer requirements, validation, and references.

In addition to such inventories, there have been a number of studies on the documentation necessary for computer models, (see, for instance: House and McLeod, 1977; Gass, 1978). Their basic suggestion, without reference to the computerized implementation of the model base, is to structure the information on each model into logical blocks, each containing items which are closely related to each other. Elaborating on their suggestions, a modelbase has been developed at the Laboratorio di Informatica Territoriale e Ambientale (LITA)(Research Centre for Computers in the

Environment), Department of Electronics, Politecnico di Milano based on the following structure for each model (Guariso and Werthner 1986):

— a descriptive block containing the name of the model, the name and address of the developing institution and an address for further reference,
— a block on model objectives and relations with other models,
— a block on techniques, methods, and assumptions used,
— a block on computer implementation,
— a block on documented applications, and
— a block on tools, support and cost to transfer the model to a different computer and/or application.

This architecture has been implemented in a PC network environment, using a commercial DBMS which also allows the direct use of some models from within the system.

4.1.2 A modelbase for small user groups

When the computer system is to be used by a specific set of users, the approach to structuring the model base may be completely different.

For instance, Guariso *et al.* (1988) have presented a software package, the Groundwater Manager's Toolkit (GMT), which is dedicated to the manager of a specified aquifer and is implemented in a PC environment. GMT provides both a set of basic tools commonly used to analyse groundwater problems, and a flexible environment in which even complex models, developed specifically for the user's situation, can be easily integrated and effectively utilized.

The core of the GMT package includes a shell written in C programming language, which is the only part of the system permanently resident in the computer memory, and twelve other programs, written in different computer languages; each of them is devoted to a specific function such as data input, aquifer simulation, pollution analysis, or display of results.

The shell controls the interaction between the user and the set of programs in such a way that they appear as different options of a large single task. However, only one program at a time is loaded into the computer memory, and the communications between the various parts take place through files, according to the scheme shown in Fig. 4.1. In some ways this structure is similar to those of well-known PC software environments such as MS-Windows† or GEM‡.

This structure has a number of distinct advantages. Different computer languages (Pascal, Fortran, C, Assembler, etc.) can be easily combined, since each program works independently of all the others. Each program can use almost all the computer memory to perform a particular task, and the structure of the complete package can thus grow to a remarkable complexity, while still running on a Personal Computer processor. Moreover,

† MS-Windows is a trademark of Microsoft Corp.
‡ GEM is a trademark of Digital Research Inc.

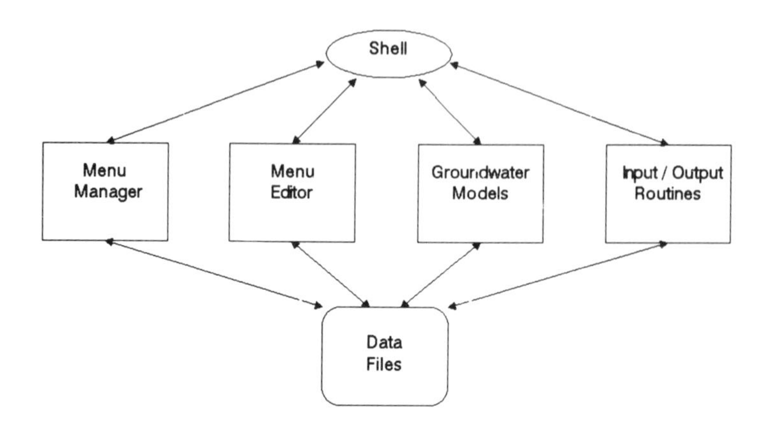

Fig. 4.1 — The architecture of the GMT software environment.

the inherent flexibility of the architecture allows a new model to be introduced into the system without perturbing the existent environment. The user's view of the modelbase is in fact entirely managed by a very simple program (Menu Editor), which reads and writes a file containing a record for each groundwater model available in the system. Such a record is divided into two fields, the first containing a short model description, which will appear in the model menu, and the second field with the program name, which is invoked at each choice. The menu editor displays the present status of the file and the user can add or delete an entry, thus enabling or disabling the choice of a particular groundwater hydraulic or water quality model. The responsibility of handling the modelbase is thus given completely to the user, who is responsible for formulating the model description that appears on the menu, which identifies the software to be executed in a form the user (or the user's colleagues) can easily interpret. Similar approaches have been used by Halpern *et al.* (1987) and Iacobucci (1985) for more general purposes. Any executable file, and thus any independently developed model, in almost any computer language can be added to the system. This modelbase structure clearly reflects a situation in which the number of models normally in use is very limited.

All the models in GMT must obviously read and write data according to a specified format. These data are either supplied by GMT input routines or displayed by the graphical tools included in the package. It should be noted, however, that a program for converting the format of data from GMT to the user's specifications or vice versa can also be inserted as a separate choice in the model menu.

Other modules in GMT are (see again Fig. 4.1) a menu manager, which displays the menus of the system and returns the user's choice to the shell for the required execution; the modelbase, which contains the groundwater models currently available within the system; and a number of programs for data acquisition and display.

Such a structure is completely transparent to the user. The software, as seen by the user, is represented in Fig. 4.2; from the main menu, the user can access the menu editor, the different groundwater models, the data-acquisition and result-presentation routines.

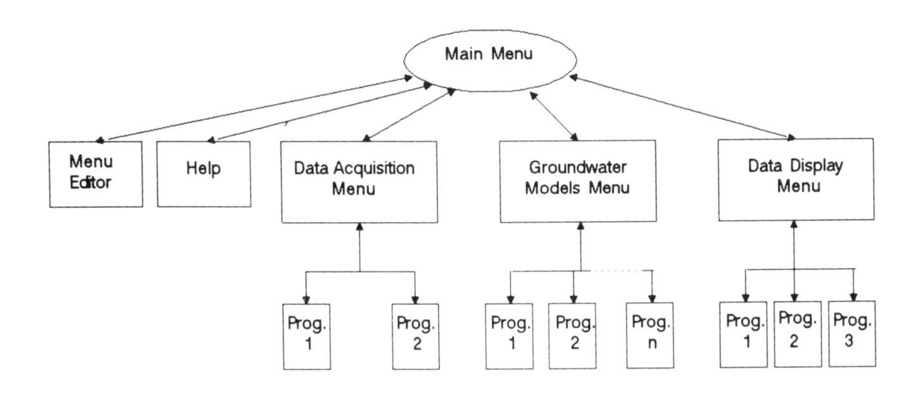

Fig. 4.2 — User's view of the GMT package: each block is a separate executable program.

4.2 TYPES OF ENVIRONMENTAL MODELS

The subdivision of environmental models based on their scope, i.e. simulation, forecast, planning, and management as presented in Chapter 2, is of little use for practical purposes. The solution of real problems requires a combination of several different models belonging to different categories. This will become clear from the analysis of the case studies which are presented at the end of this chapter. The purpose of the present section, however, is to discuss some of the distinct characteristics of the various types of models.

4.2.1 Simulation models

Simulation models try to emulate on a computer the behaviour of a real system with all the facets presented in section 2.3. In this section, reference is made to models built for the specific purpose of replicating the dynamics of a system in order better to understand the structure and connections which exist inside that system. In general, this is a preliminary step, taken before a decision is made on how to build the system or to act on it. For this purpose (see below), it may still be necessary to use repeated simulations as a tool for finding an acceptable design or action. This is why some authors (Elzas 1984, Schmidt 1988) distinguish the second use as a 'system design phase' or a 'goal-oriented simulation'. In both uses, however, the general structure of the model does not change (even if a more accurate portrait of certain details may be typical of models employed to improve understanding of real systems).

Simulation models are usually characterized by the structure shown in Fig. 4.3. The input variables u represent a set of external conditions, normally varying in time; the state variables x contain all the information about the past behaviour of the system which influence the subsequent evolution, and the output variables y represent the set of values which the user can measure to detect conditions of the system. Each model of this type is thus characterized by two (vectorial) functions. The first function, called the 'state transition function', describes the evolution of the system state as a function of the previous values of the state itself and of the input; the second function, called the 'output transformation', relates the values of the output to the state.

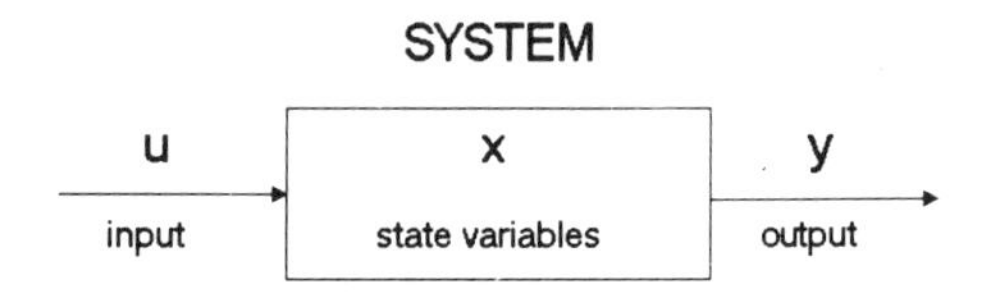

Fig. 4.3 — General structure of a simulation model.

Both of these functions may assume different mathematical forms (which are defined during the conceptualization phase), but in general they depend upon a number of unknown parameters, which must be estimated in order to fit the model to a specific real problem (see section 2.1).

In the case of continuous models, the state transition function is implicitly defined by means of a (vectorial) differential equation which relates the derivative of the state, with respect to time, to the input and previous state values. This differential equation may be based on total derivatives, if the state is represented by a finite number of values, or on partial derivatives, when the state is also a function of some other independent variables representing a spatial coordinate. For example, all diffusion and transport processes are usually described with partial differential equations.

More formally, a simulation model can thus be defined as a set M such that

$$M = (T,x,u,y,\Omega,\Gamma,\varphi,\mu)$$

where T is the set of possible values of time (a subset of real numbers) and of a time unit,

x $=(x_1,x_2,\ldots,x_n)$ is the set of state variables,

u $= (u_1,u_2,\ldots,u_m)$ is the set of input variables,

y $= (y_1,y_2,\ldots,y_p)$ is the set of output variables,

Ω is the set of possible input functions,

- Γ is the set of possible output functions,
- φ is the state transition function $\varphi(t,x,t^0,x^0,u[t^0,t),\Theta')$ which allows the state $x(t)$ at time t to be computed as a function of the initial state x^0 at time t^0, of the input in the interval $[t^0,t)$ and of a set $\Theta'=(\Theta_1,\Theta_2,\ldots,\Theta_r)$ of parameters,
- μ is the output transformation, namely a function $\mu(y,t,u,x,\Theta'')$ which allows the output values y to be computed on the basis of time, input and state variables, and a set $\Theta''=(\Theta_{r+1},\Theta_{r+2},\ldots,\Theta_q)$ of other parameters.

For a state vector **x** with a finite dimension, the most common form of the state transition function is given by a set of differential equations

$$\mathrm{d}x/\mathrm{d}t = f(t,x,u,\Theta')$$

for continuous time systems or by a set of difference equations

$$x(t+1) = f(t,x(t),u(t),\Theta')$$

for discrete time systems. Models whose state transition function does not depend explicitly upon t, which is very common in practice, are called time-invariant or stationary.

A subset of this type of model includes those which do not have a set x of state variables and thus no state transition function φ, but simply an output transformation which directly links input and output values. These models are sometimes referred to as 'memoryless' or 'static'. It should be noted, however, that memoryless models simply represent an algebraic transformation from input to output without dynamics and thus they do not catch one of the basic features of any environmental system. For this reason, they may enter into the structure of a more complex model, but they cannot represent the system to be simulated on their own.

Obviously, when time is sampled at discrete intervals, the derivatives must be replaced by unit increments and thus the differential equations by difference equations. The latter is the unique form which can be handled on a digital computer and thus, in practice, any simulation corresponds to the solution of a set of difference equations. The proper derivation of this set when the model's original formulation was continuous is the function of numerical integration, which also helps in understanding the kind of errors due to this transformation. In extreme synthesis, these errors are a function of the sampling step, i.e. of the length of the time increment considered. For very small time increments (but larger than the precision of the computer) all integration procedures obtain sufficiently accurate results at the obvious expense of the supplementary computer time required to perform a simulation of the same time horizon t^0-t. One thus has to trade off between accuracy and speed of execution, but this problem may become less and less relevant in the future, since computer speed is steadily increasing and effective integration procedures are already available also for parallel processors.

In spite of the technical difficulties mentioned above, simulation is normally an easy task to perform on any computer and this allows the modeller to take into account many of the facets of the real system. This characteristic is not shared by planning and management models, the solution of which requires, in almost all practical cases, a great degree of simplification of the problem at hand.

Another approach in simulating the time-dependent behaviour of physical systems is the so-called 'discrete event simulation'. Models of this type have the same structure with regard to input, state and output variables, but, in contrast to those presented above, the state changes only when discrete events occur. These events may be either external (in which case they can be seen as input) or internal, depending on the state of the system. Events occur at points in time, which have no duration and may not be equidistant. Between two events, the simulated system remains in the same state. This leads to a paradox at the implementation level: when the computer is working and consuming time, the simulation time does not continue (an event occurs, the computer has to calculate the new state). When the computer is not working, simulation time is passing (the model is between two events and virtual time has to pass from the first to the second event, but the computer has nothing to work on since no event has actually occurred). A very simplified version of population dynamics can be used to illustrate a real system which might be formalized by a discrete event model. If the number of living individuals is defined as the state variable, the system changes its state only with the birth or death of an individual. Births and deaths are thus events occurring as inputs at non-equidistant points on the time axis.

The use of models for simulation implies the definition of what is usually called an 'experiment' (Zeigler 1976). This means that all the variables that are not relative to the model structure have to be fixed. An experiment E is thus

$$\underline{E} = (\underline{t^0},\underline{t},\underline{x^0},\underline{u[t^0 t)},\underline{\Theta})$$

and corresponds to supply values for the initial t^0 and final time t, the initial values of the state variables x^0, the input functions u for all the interval and a set of values for all the parameters ($\underline{\Theta} = \underline{\Theta}' \cup \overline{\Theta}''$). The notation $u[t^0,t)$ refers to the fact that the input values must be fixed from the initial time to a unit time step (or an infinitesimal) before the final time, since the value of $x(t)$ will be determined by the dynamics of the system. In an EDSS, an experiment is thus performed by selecting a model from the modelbase, accessing the required data from the database, and possibly prompting the user to supply the other necessary information.

A careful definition of experiments is essential to obtain informative results from the simulation model. In particular, the parameter values $\underline{\Theta}$ may correspond to the results of the calibration, and the final results are rather insensitive to the initial condition x^0 since the simulation is normally run on a time horizon which is very long with respect to the interval in which the output is driven by the initial values of the state. In contrast, the choice of the input values is always critical. Without dealing with theoretical considerations about the relationship between model structure and meaningful input values, which have been mainly developed for linear models (see, for instance, Luenberger 1979), in real cases, it is normal that input values do not depend (or depend only partially) upon the decisions which can be taken by the person (or agency) who runs the simulation. Input variables may in fact be somebody else's responsibility or be affected by natural, sometimes random, external facts. The closer these external inputs are to the real scenario in which the system will evolve, the more informative will be the results obtained by the simulation model.

There are a few ways to handle this situation. The most common is to assume, as values for the external input, those actually recorded on the system in the past. In this case, the results of the simulation can be easily compared with data recorded on the real system. Another approach is to estimate the statistical properties of the input variables from past records and then generate other synthetic records with the same statistical characteristics. The simulation is then run with each of these input samples, and the distribution of all output values obtained in this way is analysed in order to relate its properties to those of the input.

4.2.2 Forecast models

A simulation model can also be used for forecast purposes if the time horizon is so limited that the effect on the output of the information (state and input values) available at a given instant of time (when the forecast is issued) is still significant. For longer time horizons, the output behaviour becomes dependent on future values of the input, which are unknown, and thus the model cannot be applied. A reasonable length of the forecast horizon may thus be set only by looking at the dynamics of the system. A fast-evolving system, for instance a small river catchment where rainfall reaches the output flow gauge within a few hours, will allow reasonable forecasts only two or three hours ahead (longer-term forecasts would entail knowledge of future precipitation), while a slower system, for instance an unexploited forest, may allow future tree biomass to be predicted several years in advance with a degree of precision comparable to the other example.

Particularly when the purpose of the model is to forecast and thus the interest for the internal behaviour of all the model components is determined only by the required accuracy, it is common to use a so-called 'black-box' description of the system, i.e. a simple mathematical structure which captures the input–output behaviour of the system without precisely following all the cause–effect relationships which determine its dynamics. Usually these kinds of models are calibrated and tested to give good prediction in mean conditions. They are never able accurately to forecast extreme episodes, and thus they are considered too risky in certain practical cases.

There are several possible structures for these black-box models, but the most common (Box and Jenkins 1970) is a linear relation of the type:

$$y(t) = \sum_{i=1}^{p} a_i\, y(t-i) + \sum_{j=1}^{q} b_j\, u(t-j) + \sum_{l=1}^{r} c_l\, \varepsilon(t-l) + \varepsilon(t)$$

where y and u have been previously defined, a_i $(i=1,\ldots,p)$, b_j $(j=1,\ldots,q)$, and c_l $(l=1,\ldots,r)$ are the model parameters, and $\varepsilon(t)$ is a random variable with zero mean, uncorrelated with $y(t)$ and $u(t)$. These models are called ARMAX (AutoRegressive Moving Average with eXogenous inputs) and have been frequently used as predictors, assuming the expected value of $\varepsilon(t)$, namely zero, and estimating the previous values $\varepsilon(t-l)$ as the differences between real values of $y(t)$ and forecasts obtained by the model.

The decision about how to use the forecast in practice is in itself a planning problem. Given the inaccuracy mentioned above, one has to decide when an alarm must be issued on the basis of the forecast values. This decision is greatly affected by the attitude of the decision maker towards risk and by the expected costs entailed in a 'false alarm'. Certain costs are incurred if an alarm is given needlessly (for instance for preparing protective structures), while not issuing a necessary alarm may result in damages and much higher costs. The trade-off between these two costs and the accuracy of the forecasting model must suggest to the decision maker which are the model results for which the alarm must be issued.

4.2.3 Planning models

Planning (or at least approving plans) has been one of the major activities of most environmental agencies. A planning decision might involve for instance the location of a plant, the height of a dam, the number of personnel involved in a public service, or which law to enforce or which taxes should be paid for using an environmental resource. These kinds of decisions can usually be translated into a mathematical programming problem, i.e. into a model which sets the values of the decisions to be taken in order to optimize certain decision criteria satisfying a set of physical, legal or economic constraints. From a formal point of view, the problem may written as

$$\max_{\mathbf{z}} \text{ (or min) } \mathbf{f}(\mathbf{z})$$

subject to

$$\mathbf{z} \in Z$$

where $\mathbf{z}$ is a vector of decision variables, $\mathbf{f}(\mathbf{z})$ is a vector of functions and Z is the feasible set of values for $\mathbf{z}$ defined by the constraints.

A more precise definition and solution of these problems falls into the field of operations research, which will not be dealt with here. The remainder of this section is an analysis of the different components of the planning model and an assessment of their formulation and role in the case of environmental decisions.

Decision variables

These are the independent variables of the problem, which must be defined in the solution procedure. They are normally identified in the formulation phase, since an agency knows what can be changed under its responsibility and what is outside its power (and thus must enter the model as an external input or a parameter value). Depending on the kind of problem at hand, the decision variables may be represented by real numbers (for instance the height of a dam) or by integers (for instance the number of plants or instruments to install) or by binary numbers (the decision to allow a new user to enter an environmental system may be just 'yes' or 'no', the variable representing these decisions being 1 or 0).

Constraints

When taking some decision on the environment, one usually has to satisfy a set of conditions which may be translated into a set of equations linking together the decision variables and other inputs and parameters. Examples of these conditions might be as follows: the total expenditure for a certain project must not exceed the budget; the pollution generated by a plant must not exceed the limits set by law; or the waiting time for a public transport system must not exceed a given threshold.

Other constraints that are considered mandatory for the performance of the system may be set by the decision maker. In this sense, constraints may represent limit values of certain other objectives that, though not explicitly considered for optimization, must not be ignored.

In technical terms, the constraints are expressed by a set of equalities or inequalities

$$h(z) = 0$$
$$g(z) \leqslant 0$$

the solution of which defines the feasible set of solutions, i.e. all the combinations of values of the decision variables are only a subset of the *a priori* possible combinations. A narrow feasibility set normally worsens the value of the optimality criteria, and a solution may not even exist if two constraints are conflicting. Thus, there must be a trade-off between the absolute performances expressed as constraints and the optimal values of the criteria; however, in many practical cases, the constraints are defined by an external situation, which is well known to the decision makers. In particular, the definition of a dynamic model, which represents the natural component of the system to plan, is one of the constraints in almost all environmental planning problems.

The optimality criteria

The most difficult question in many environmental decision models is how to formalize the criteria which must guide the choices. In this connection, there has been a change in attitude in the last decades (Kindler 1988), which is briefly summarized below.

In 1936, the US Flood Control Act stated for the first time in environmental planning that, in any project to be considered, 'benefits to whomsoever they accrue must exceed costs' or, using more modern terminology, a project had to have a cost–benefit ratio lower than one. This introduced cost–benefit analysis as a formal criterion in considering environmental projects and generated an amazing number of publications on this topic (see for instance Marglin 1962, Herfindhal and Kneese 1974, Peskin and Seskin 1975, Fronza and Garofalo 1979).

The formulation of a decision criterion in terms of the maximum difference between benefit and cost appealed to economists, who felt that the primary role of their discipline in environmental planning had been recognized, and to modellers, who felt themselves well equipped for solving problems by translating the decision-maker's criteria in a unique and objective manner. Provided, in fact, that benefit and cost could be expressed as an analytical function of the decision variables, there were plenty of methods available to reach at least a suboptimal solution of this single objective problem.

This generated a rush to develop analytical formulas to assess the costs and benefits of all facets of any environmental problem. However, the cost–benefit approach failed almost completely in the sense that very few decisions have been taken only on the basis of its results. Two major theoretical problems have prevented a more widely accepted use of the method. The first is the fact that some benefits and costs of environmental plans are difficult to measure, especially in monetary terms, such as a beautiful landscape or the cost of a human life, while others are only indirect, the increase in tourism due to planting new trees in a forest depleted for some external causes. The second problem was that benefits and costs accruing to different social groups could not easily be added together. The pollution generated by a plant obviously represents a certain 'benefit' to the plant owner (more pollution means more production and saving of cleaning costs) but it worsens the living condition of people even far away from the plant, who may have extra production costs in order to use the same resource (for instance, more filtering of the air for pharmaceutical or electronic industries) or extra medical costs to preserve their health. These two groups, polluters and damaged, may even live in different countries and direct compensation in monetary terms is almost always an unacceptable solution.

The recognition of the complexity of all environmental problems and of the virtual impossibility of reducing all their facets to an algebraic sum of benefits and costs led the way, in the second half of the seventies, to the application of another methodology which allowed all of the objectives of the decision makers to be considered separately. Researchers thus turned their attention to multiobjective programming and another huge amount of literature was published on the subject (see for instance Cohon and Marks 1975, Hall *et al.* 1975, Major 1977, Cohon 1978).

The rationale behind this method is very simple. Each aspect of the problem or point of view of the decision maker is translated into a separate objective function which uses a specific and independent unit. For instance, one function may measure the construction costs in dollars, another the pollution in milligrams per litre and a third the electricity production in megawatt-hours. From a technical point of view, the scalar optimization (which normally allows the determination of a unique optimal solution) became a vector optimization, which does not result in a unique decision.

The 'optimality' of a vector can in fact only be defined using the old formulation originally developed by the economist Pareto at the beginning of the century. In this context, a solution (i.e. a decision) A is said to (strictly) dominate another solution B if and only if A corresponds to values of all objective functions which are preferable to those due to B. It is clear that this allows all the dominated solutions to be disregarded as non-optimal, but still may leave more than one non-dominated (or efficient or optimal) solution. If the feasible alternatives are numerous, the number of non-dominated solutions may be also very large and infinite at the limit (this happens if one or more decision variables are continuous).

Although this approach may be useful in discarding alternatives which are not worth considering, it leaves the final decision problem, i.e. arriving at a single solution, unsolved. In particular, when there are strongly conflicting points of view, representing for instance the interests of different social groups, the Pareto optimality may not give a significant result since even the extreme solutions, i.e. to give all the resources to one group or to the other, are by definition non-dominated.

The method thus produces some practical results only with very few objectives (two or three) and particular shapes of the non-dominated set of solutions that allow an acceptable compromise to be easily established between the various points of view.

In the last few years, another method has become more and more popular among environmental decision makers. Called 'Environmental Impact Assessment' (EIA), it has been applied in several different ways in different countries and environmental sectors (Lee 1983) and is based on a set of commonly accepted principles. First, the real alternative solutions in many environmental plans are limited in number. Second, each alternative should be analysed according to a large set of different points of view. This approach thus differs considerably from the previous ones. Instead of defining implicitly the solution as the decision which optimizes the objective functions, it starts directly from a solution and computes all its effects on the environment. Each feasible plan thus has a vector of values which indicate all its impacts. These values may be the output of simulation models which portray a particular sector or may even be qualitative values assigned by a group of experts using some kind of 'mental model' of a facet which is particularly difficult to measure in quantitative terms.

At the end of this process, the decision maker is still confronted with the problem of choosing one alternative, but here the choice is normally confined to a very limited number of solutions. Some guidelines can be used in making that choice. First, solutions which violate explicit environmental standards or an acceptable level of impact must be discarded as well as dominated solutions. A common approach used to choose among the remaining alternatives is to set a weight for the relative importance of each computed impact and to rank the alternatives by computing for each of them an aggregate performance index as the weighted sum of the impacts. A great deal has been written on determining appropriate weights (see Keeney and Raiffa 1976, Krzysztofowicz and Duckstein 1979), and so this will not be discussed here. In principle, however, weighting is a political decision and thus can only be carried out by decision makers, who in turn may directly consult with the social groups involved in the real system.

However, even for a perfectly independent and objective decision maker, it may be extremely difficult to set all the weights in a consistent way (Inhaber 1976). It is usually easier to make a sound comparison between only two criteria and then find, by means of a mathematical program, the consistent set of weights that is closest to the relative weights expressed by the decision maker for each pair of criteria.

Another idea (Colorni and Laniado 1987a) is to start from an initial configuration of the weights (and thus from a certain ranking of the alternatives) and explore which variations must occur in the weights in order to change the ranking. For instance, if one solution is preferred to another for any value of the weights between 1 and 100, there are good chances that the second solution can be discarded from further analysis. After discarding some alternatives the weights may be recomputed and the process repeated.

None of these ideas obviously 'solve' the environmental planning problem, which remains, as it should, a subject of political debate. However, it is clear from the current trends that the idea of an 'optimal' plan is being replaced by the idea of an 'acceptable' plan, and this increases the role of simulation models, which must compute all the consequences of alternative decisions in order to check their acceptability. The role of optimization models, at least until a more powerful conceptual framework is proposed, will be limited to certain specific subproblems where one or very few predominant criteria can be easily formalized and accepted by all parties involved.

4.2.4 Management (or control) models

The problem of managing an environmental system can be formalized in more or less the same way as in the planning case, except that the optimization function depends explicitly on the behaviour of the system on the time horizon under consideration. Formally, this problem can be stated as

$$\max \text{ (or min) } \int_{t^0}^{t} f(\tau,x,y,\Theta,u,z)\, d\tau$$

where x,y and Θ are the state, and output variables and parameters of the system, u represents the set of input variables which can be fixed by the manager and z another set of input variables outside the manager's control. This objective is subject to a set of constraints that includes the state transition function and the output transformation of the system. The solution to this problem is a management (or control) policy $u^0(t,y,z)$, which allows the input at each time to be determined on the basis of the available information.

All of the problems discussed in the section about planning and related to the multiple objectives of the decision are still valid. However, the objective of the management problem is sometimes simpler to formalize, since often a previous planning problem has already been solved and the desired system behaviour has already been fixed. The management problem can thus be considered as a single-objective one: namely, following the planned pattern as closely as possible. In this sense, control problems appear to be characterized by a more defined structure than planning problems, as was already pointed out in Chapter 1. However, this does not mean that these models will not be present in an EDSS modelbase since, the solution of many planning problems involves a hypothesis about the management of certain subsystems.

Even if the objective is more clear than in planning problems, managing is quite a difficult task since a decision taken at a certain time influences the system evolution and thus all subsequent decisions. Only in a very few, almost ideal cases (such as a linear system with a quadratic optimality criterion) are theoretical results available. Under all other conditions, the control problem must be solved by empirical procedures which cannot guarantee the optimality of the decision. Furthermore, only experience gained on a real system can give an indication of how far from optimal the proposed solution is.

In management (or control) problems, a critical role is played by the information structure of the system. While solving a planning problem all the information related to the entire planning horizon must be available at the time the decision is taken, a management decision can be made at each time interval, only on the basis of the information available up to that time. Obviously, one may wish to collect all possible data on a continuous basis, but this means spending money on a measurement network which may not result in a significant improvement in the control performance. Thus, the information system must be planned with a view to optimizing the difference between the benefits of a better control decision and the costs of the information system. Again, this means that, before solving a management problem, one should in principle solve the planning problem of information collection.

Control theory has illustrated the relative advantages of different control structures, which are briefly summarized below. The minimum information needed to manage a periodic system (such as an environmental system, see section 2.3) is time. In this case, the scheme is reported in Fig. 4.4 and the

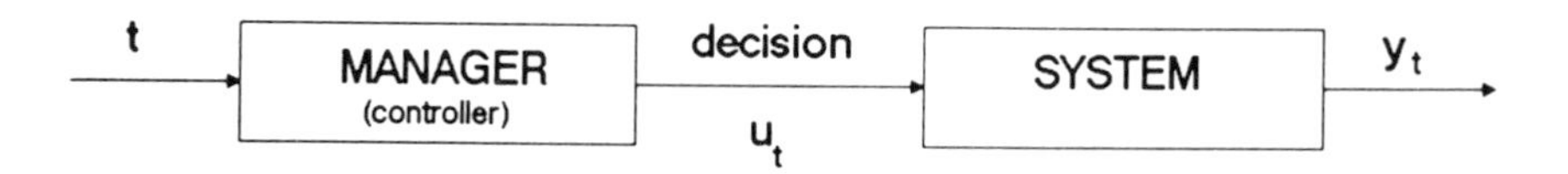

Fig. 4.4 — Open loop control structure.

decision is taken only on the basis of the day of the year or the hour of the day and thus $u^0 = u^0(t)$. This 'open-loop' management is clearly very cheap (no measurement is taken), but very weak in the sense that any deviation of the system from its standard conditions, owing to a deviation from nominal values of the inputs which do not depend upon the manager, may bring the system very far from the desired performance.

A more effective information structure ('closed loop' or 'feedback') is represented in Fig. 4.5. In this case, the manager also receives data on the

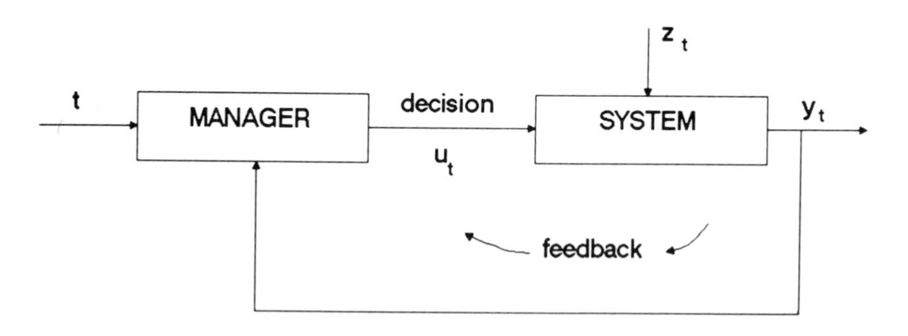

Fig. 4.5 — Closed loop control structure.

actual behaviour of the system ($u^0 = u^0(t,y)$) and thus can recover more easily from undesirable deviations. The counterpart is the cost of measuring the system behaviour and of developing a more complex control rule.

The preceding structure can be further improved by providing the manager with external inputs z or even with the causes which produced those inputs ('feed-forward loop'). The structure derived in this way (see Fig. 4.6) involves a larger number of measurements and thus higher costs and a more complex control procedure based on a wide number of variables $u^0 = u^0(t,y,z)$.

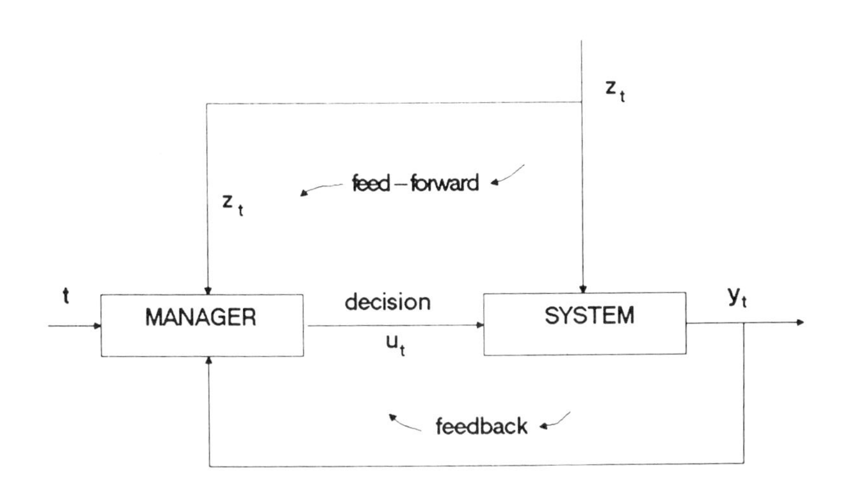

Fig. 4.6 — Feedback and feed-forward control structure.

A specification of the latter type of structure is obtained by dividing the controller into two parts: the first is a forecast model which computes, from available measurements, the future external inputs to the system; and the second is the real control rule, which is also based on the forecast values (Fig. 4.7). An application of this procedure which mixes forecasting and management models, will be discussed in detail at the end of the chapter.

Managers of real systems usually use as much as possible an information structure of the latter type; however, they process all available data in an implicit and subjective way, based on their experience on the system, and it is often quite hard to synthesize this way of thinking into a formal procedure. Attempts have been made, in the case of reservoir management, by using expert systems (see section 2.6) and by finding a control rule which approximates as closely as possible the recorded behaviour of an existing manager (see Zielinski *et al.* 1981, Guariso *et al.* 1986).

This approach shares, with others (Houck *et al.* 1980), an interesting feature which makes the solution it suggests more acceptable to managers. In practice, the solution of the management problem, which is the function $u(\cdot)$ that links the decision to the available information, is limited *a priori* to a fixed class of functions $u = \Gamma(t,y,z,\Theta')$. In this way, the solution can be

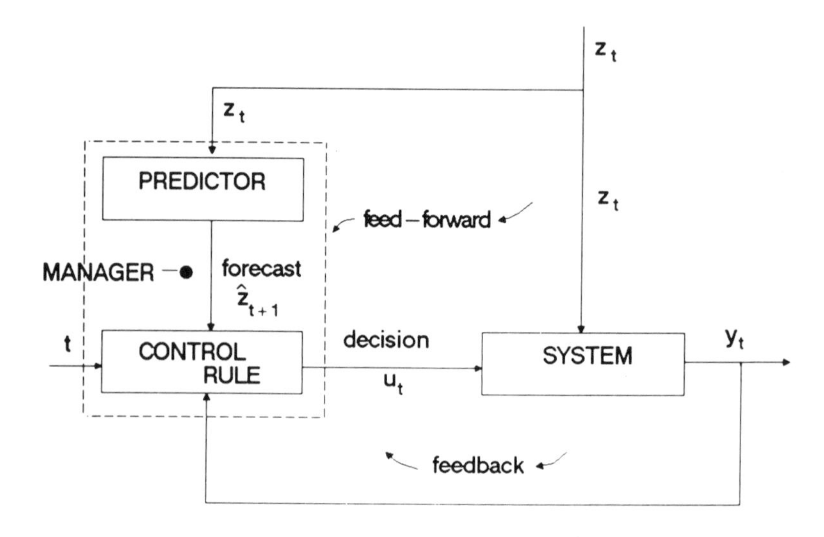

Fig. 4.7 — A control structure using a forecast model.

found by solving a mathematical programming problem, which determines only the values of a few parameters Θ' defining the optimal function within the class. Furthermore, having fixed *a priori* the class of control functions means that the solution cannot show strange shapes, which, even if technically optimal, cannot be accepted by managers. Other minor or implicit constraints, approximations in the problem formulation or unforeseen deviations of external input are usually considered more significant in practice than some suboptimality of the solution.

For the same reason, formal optimal procedures, such as the maximum principle of Pontriagyn or various versions of the dynamic programming algorithm of Bellman, have also been tried on several occasions on environmental systems (see for instance Loucks *et al.* 1981, Yeh 1985) but failed to generate acceptable solutions. Their optimal solution frequently corresponds to an extreme behaviour which the manager does not like to implement.

This draws attention to the 'risk-averse' attitude that all managers have shown. This is not hard to understand; if a decision has been taken in the past and has been accepted, there is a certain margin of risk in changing. This attitude may be partially portrayed in problem formulation (see for instance McBean *et al.* 1978) by forcing the solution to behave in an acceptable way, even under extreme conditions, or by exactly optimizing the system behaviour in those conditions. This is a common practice among environmental engineers who have designed management rules to withstand episodes which happen only once a century, or every thousand years, or are the worst in recorded history. The safer the rule must be, the more conservative it is; and thus in this case the attitude towards risk must be traded off against the implementation costs.

If one optimizes the management in a suitably defined 'worst' situation, some freedom is usually left when the situation is less severe. It is thus possible to exploit this freedom in order to optimize another objective with the constraint of always guaranteeing the same performance in difficult cases. Two optimization problems may thus be solved in cascade. First, the most important objective is satisfied, and second, some less important index is improved. If the solution of the second problem is not unique, one may optimize a third problem without losing anything on the first two and so on. In practice, if the worst case considered is really rare, there is plenty of freedom most of the time, and in some cases even the second objective may be taken very close to its independent optimum (see Gandolfi *et al.* 1988).

4.3 A COMMON MODELLING FRAMEWORK

As emerged from the previous discussion, the solution of real problems normally requires the use of several different types of models. To calibrate a simulation model, one has to use an optimization model; to plan a system it may be necessary to consider how it will be managed; and to control a plant may imply forecasting future conditions. It is thus of relevance to find a common framework for integrating all these different types of models into an EDSS modelbase. Only few attempts to do so have been presented in the literature and they were mainly aimed at constructing simulation software packages, including some parameter estimation facilities (Birta 1984, De Buyser and Spriet 1988, Ruzicka 1988).

For this purpose, a multiple experiment can be introduced as an extension of the (elementary) experiment previously defined. A multiple experiment ε is defined as a set $(E_1,E_2,\ldots,E_N)$ of experiments and a method δ for performing those experiments. A method may be, for instance, to access the database several times to get different values of the input functions. A slightly more complex method may be to subdivide a range of possible parameter values into evenly spaced intervals and simulate the system in correspondence with those values of the parameters. A third, still more complex, method may be to plan new experiments on the basis of the results of preceding ones, for instance, with the aim of optimizing a suitable performance indicator. The latter is particularly important because it allows a planning or management model to be interpreted as a multiple experiment performed on a simulation model.

To better define a method δ, one may characterize it by the following:

- a maximum number N of elementary experiments to perform
- a range of variation (or a set of values) for Θ and/or x^0 and/or u
- an (optional) function of $\beta(t,u,x,y)$ to evaluate
- an (optional) set of constraints on x and y
- an (optional) procedure to compute new values for Θ and/or $x^0$and/or u on the basis of preceding results of β.

It should be noted that this definition of the multiple experiments includes several different operations which are normally performed with models. Parameter estimation is simply an optimization procedure, where parameters are the decision variables and the function to optimize is a measure of the departure of the model from the real system. Sensitivity analysis means evaluating the deviations of states or outputs from a given behaviour in correspondence with a variation of the input or initial conditions or parameter values.

The set of constraints mentioned above simply has the role of stopping any elementary experiment where some undesired or unfeasible situation occurs. Such constraint could be included in the definition of an elementary experiment as well, but in the case of a single run, one may easily detect the violation of these constraints and only when trying several different elementary experiments they may save a significant amount of computer time.

The procedure to set new values of the unknown independent variables (Θ, u, x^0) on the basis of the results of previous experiments may be constituted by any operations research algorithm. In general, its choice may be made by the user, but a set of rules may be present in the knowledge base of an EDSS to suggest a suitable procedure based on the number of values to optimize, certain structural properties of the model, and the number of elementary experiments (and thus the computer time) the user has decided to spend.

Such a procedure may be stored either in the model or in the knowledge base. From a representation point of view, since these methods may be described in a procedural form similar to simulation models, they may enter the modelbase as separate items. In this case, the user will be able to choose a simulation or an optimization model in much the same way. From a functional point of view, however, such methods describe how to use other models, and thus can be stored in the knowledge base with other informations of that type. This separation between models of real systems stored in the modelbase, and methods of using them stored in the knowledge base, may be a useful guide in the construction of an EDSS, which is demonstrated in Chapter 7.

4.4 EXAMPLES OF WORKING SYSTEMS

This section analyses some computer programs which, while not exactly fitting into the definition of an EDSS presented in chapter 2, may be helpful in solving common environmental problems. In general, their limitation is the very narrow problem domain they address. Sometimes they are general packages for a certain class of systems, like WODA, while others, like VERBASIM, are devoted to the study of a unique case. Though supplying the user with a series of different models, they do not allow a straightforward insertion of new models, nor do they permit existing ones to be connected in different ways.

Although computer packages of this kind are becoming more and more common in environmental studies (see, for instance, Fedra 1985, Guariso *et al.* 1985b, Loucks *et al.* 1987, Golden *et al.* 1987), the examples presented here share common features which are of interest in the present study: they were developed under the pressure of real problems (and not academic ones) and have been in operation for a few years, so that a preliminary balance of experience can be attempted.

4.4.1 WODA: Water Oxygenation Deoxygenation Analysis

WODA, the Polish term for water, is a computer package developed by Kraszewski and Soncini-Sessa (1984, 1986) to assess the impact of different patterns of biodegradable discharges on the quality of a river. The package is thus meant to assist a river agency in taking decisions on, for example, the level of treatment required by a certain discharge point, the installation of artificial aerators, or the application of a control measure to limit the runoff of biodegradable substances from agricultural areas. All these factors cause a variation in the pollution of a river and may alter its oxygen content at some downstream point. The degradation process in a river can in fact be approximated, as first proposed by Streeter and Phelps (1925), as a chemical reaction between dissolved oxygen (DO) and a pollutant, measured in terms of the oxygen used by bacteria to degrade it or biochemical oxygen demand (BOD) (see for instance Rinaldi *et al.* 1979, for a detailed discussion). An excessive discharge of pollution at one point may cause downstream river oxygen to reach such low values that fish cannot survive and water cannot be used for other purposes (drinking, etc.).

WODA is targeted to an audience of environmental engineers who lack experience in using mathematical models and computers and it thus assists them in all the main phases of model implementation: data collection, parameter estimation and model use. For the purpose of illustration, its structure may be subdivided into database and modelbase. The user interface is not unique, however, but differs from one base to the other. It will thus be illustrated together with the relative base.

Only the PC version of WODA will be described here. It is constituted by a master program written in C language, which calls the various modules, written in other programming languages.

WODA database

The estimation of parameters of any water quality model requires a relevant number of observations. These observations cannot be made at any time, but must follow a precise schedule in order to supply the model with the necessary information. To allow the use of mathematical methods which greatly simplify the general system of partial differential equations representing BOD and DO reaction and transport, data on both oxygen and pollutants must be collected in such a way that they represent the inputs seen by an observer moving downstream at the same velocity as the flow. Under this hypothesis, in fact, the partial differential equations may be rewritten as a system of total differential equations.

The database thus has the function of reminding the user of this method of data collection; allowing storage and retrieval of a data collection campaign; helping in the selection of a consistent set of data; and allowing separate data items to be edited. Furthermore, the DBMS permits reports and simple statistical evaluations of the data to be prepared.

The database has been implemented using a commercial relational DBMS (DBIII)† and is completely menu-driven. Each tuple is formed by fields containing either a geographical description of the river stretch under examination (name, length, distance from the river outlet, names of the measurement points, their position and so forth) or data on each collected water sample (name of the station, date, temperature and chemical composition of the water, etc.). All data are entered via simple fill-in forms (a sample is presented in section 6.6), which are also used for output.

WODA modelbase

The modelbase of this package contains a unique class of models, i.e. an extension of the water quality model proposed by Dobbins (1964), which in turn has improved the model of Streeter and Phelps (1925). The model can be written, using the hypothesis already mentioned of an observer moving at flow velocity, as

$$\frac{\mathrm{dBOD}}{\mathrm{d}t} = -k_{\mathrm{b}}\,\mathrm{BOD} + b_{\mathrm{p}} + b_{\mathrm{np}}$$

$$\frac{\mathrm{dDO}}{\mathrm{d}t} = -k_{\mathrm{d}}\,\mathrm{BOD} + k_{\mathrm{r}}\,(\mathrm{DO}_{\mathrm{s}} - \mathrm{DO}) + k_{\mathrm{a}} + d_{\mathrm{p}} + d_{\mathrm{np}}$$

where BOD and DO represent the concentration of pollutant and oxygen, k_{b}, k_{d}, k_{r} and k_{a} are parameters (known respectively as BOD decay rate, deoxygenation rate, oxygen reaeration rate and net algal photosynthetic production rate), DO_{s} is the oxygen concentration at saturation, and d_{p}, d_{np} and b_{p}, b_{np} are the inflows of oxygen and pollutant (separated in point and non-point sources) from outside the river.

The four parameters k_{b}, k_{d}, k_{r} and k_{a} are in turn functions of the temperature T and of the flow rate q, which may vary from one season to another and from one stretch of the river to another. WODA contains the hypothesis that the dependence may be written as:

$$k_{\mathrm{b}} = k_1\, k_2^{(T-20)} + k_3$$

$$k_{\mathrm{d}} = k_1\, k_2^{(T-20)}$$

$$k_{\mathrm{r}} = k_4 k_5^{(T-20)} + q^{k_6}$$

$$k_{\mathrm{a}} = k_7 + k_8\, P(t)$$

where $P(t)$ is a periodic function representing the daily distribution of

† Dbax (DBIII) is a registered trademark of Ashon-Tate.

photosynthetic activity, i.e. the number of hours of light during each day of the year.

The complete description of the degradation of a pollutant according to this kind of model implies a knowledge of the values of all eight parameters k_i, $i = 1,\dots,8$ in each stretch of the river in which temperature and flow may be considered as constant.

The model can be simplified if the parameters do not depend upon T and q (it corresponds in this case to Dobbins's model) and if pollutant sedimentation and algae are disregarded, i.e. $k_d = k_b$ and $k_a = 0$ (which represents the original Streeter-Phelps formulation). Thus, one may also see the modelbase as containing three different types of models from which the user may choose.

An experiment with a model of this type is thus defined by a pattern of flow and temperature along the river, a set of eight parameters for each stretch of the river, with constant flow and temperature; BOD and DO conditions at the headwaters; the position and flow of concentrated BOD and DO inputs; and a pattern of distributed BOD and DO inflows. It is clear that the definition of all these values is a rather time-consuming task, and thus the modelbase can automatically extract from the database all the required information once the date of the data collection campaign on the river at hand has been specified.

A graphic interface and a simple command language allow the user to change any of the values mentioned above and perform the experiment by simulating the system of differential equations given above (Fig. 4.8). In this particular case, the simulation does not require the discretization of the differential equations, since the analytical solution of the system is known and can thus be computed in one step for any particular position along the river. BOD and DO concentrations, flow and saturation values are plotted on the screen together with a graph representing the position of all the discharge points.

A last feature of WODA is that it allows the definition of the distributed BOD inflow (b_{np}), which is usually very hard to measure, as a pattern defined by the user times a parameter k_9. The pattern is usually assumed to be the distribution of equivalent inhabitants along the river (a number much easier to ascertain) and the parameter k_9 may be estimated together with the other eight using a parameter estimation procedure.

WODA provides some knowledge for estimating the parameters on a given river system. However, since it is embedded into the package and not stored in a separate module, it cannot be considered as a true knowledge base. The user may perform the estimation by running several experiments with different values of the parameters or let the system plan a multiple experiment to evaluate the best values of all k_i or of only a subset of the k_i. The estimation may be accomplished with different methods and driven by a variety of objective functions. The user may choose the latter as a weighted sum of the squares of errors on BOD and DO measurements, namely as

$$\min_{\{k_i\}} [\alpha \Sigma_t (\mathrm{BOD}_t - \underline{\mathrm{BOD}_t})^2 + (1-\alpha)\, \Sigma_t (\mathrm{DO}_t - \underline{\mathrm{DO}_t})^2]$$

were measured values are underlined.

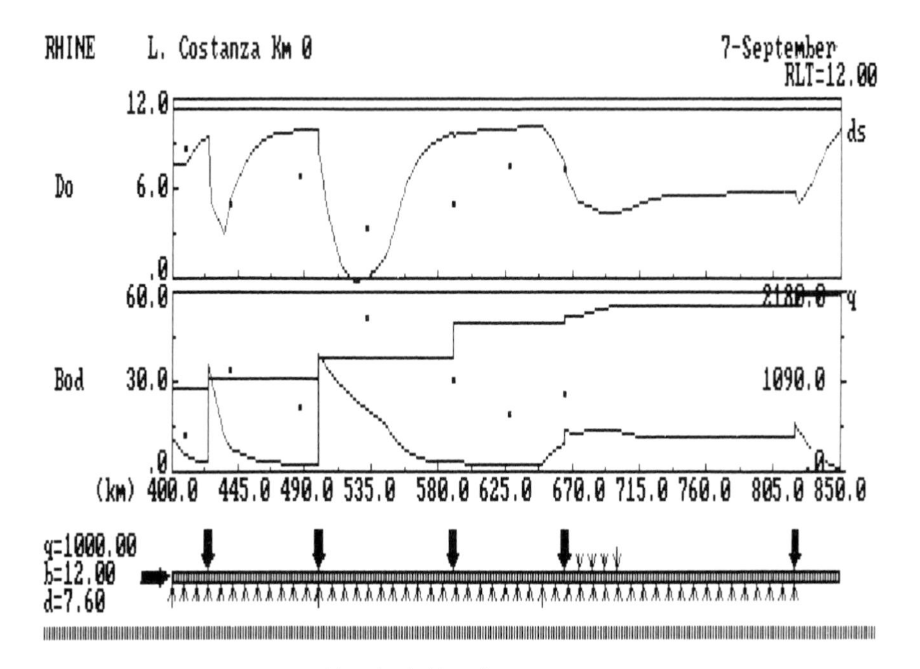

Fig. 4.8 — Simulation of BOD and DO loads on the River Rhine.

Selecting a value for α such that $0 \leqslant \alpha \leqslant 1$, the user may decide to find a model that is more precise on BOD ($\alpha = 1$) or DO ($\alpha = 0$) or something in between. This option is of particular interest to environmental engineers, who are well aware of the much higher precision of oxygen measurements with respect to BOD ones.

While the user may have a feeling for the best type of criteria to use for estimation, he or she will probably ignore the estimation techniques almost completely; thus, WODA has a built-in, very simple rule-based system which enables the user to choose from among a set of available procedures. For instance, if the parameters are constant within the river stretch under analysis, then a method proposed by Rinaldi *et al.* (1977) is used, thus considerably reducing the time required. If that assumption does not hold, a general pattern-search method (Zangwill 1967) is applied. Both the algorithms are again modified if the user decides to fix the value of certain parameters and optimize only the remaining values.

The knowledge part of WODA contains allowable limits for all the variables, checks if the computed oxygen content is always positive (in anaerobic conditions, i.e. when the DO concentration reaches zero, the model loses any validity) and warns the user in the opposite case. It will probably be enriched in the future with a more extended rule base to suggest suitable default values of all the parameters on the basis of some very simple characteristics of the river, as has already been done for the water quality model QUAL2E (Barnwell *et al.* 1986).

Outcome of the experience

WODA has been distributed to about 50 environmental agencies and consulting firms, together with a three-day course to introduce the model in some detail. Applications have been reported for several different rivers in very different conditions (Ticino and Garza rivers in Italy, Skrwa river in Poland); all the users have agreed on the usefulness of the package for their environmental planning problems.

Very few criticisms of the user interface and the program structure in general have emerged; even after the three-day course, the major difficulties arise in understanding the model's behaviour. Experience has shown that many environmental engineers lack the ability to follow the effects of a pollution load in quantitative terms. One problem may be that they start using WODA immediately after running the test case to solve a pending problem without taking the necessary time to become familiar with the model and its behaviour. This seems to indicate that it is necessary to enrich the knowledge part of WODA with some facilities to explain the results and suggestions on the use of the model.

4.4.2 VERBASIM: planning the structures and the management of a large reservoir

VERBASIM (the acronym comes from the words 'Verbano', the ancient name of Lake Maggiore and 'simulation') is a computer package (De Simoni *et al.* 1986) designed to assist the managers of Lake Maggiore in replanning its large catchment between Italy and Switzerland. VERBASIM does not represent a general EDSS as outlined in the first two chapters, nor a general package as presented above. However, it was designed to solve a very complex (and thus inherently unstructured) problem and has a quite rich (even if not user modifiable) modelbase.

Before presenting the structure of VERBASIM and its main feature, it is necessary to summarize briefly the physical, social, and economic characteristics of the problem it addresses.

Lake Maggiore planning problems

The Ticino river drains a catchment of 6700 square kilometres in the central part of the Alps and has been regulated since 1943 by a dam located in Sesto Calende at the outlet of Lake Maggiore (see Fig. 4.9). Its mean flow rate is about 300 cubic metres per second and it normally fills the lake twice a year as a result of snow melt and heavy autumn rains. During the rest of the year the water is spilled from the lake mainly to supply irrigation and hydropower through a complex network of canals.

Two main agricultural districts, Villoresi and East Sesia, with a total surface area of 2000 square kilometres, are irrigated by Lake Maggiore waters (see Fig. 4.10). Their canals can distribute water anywhere on the

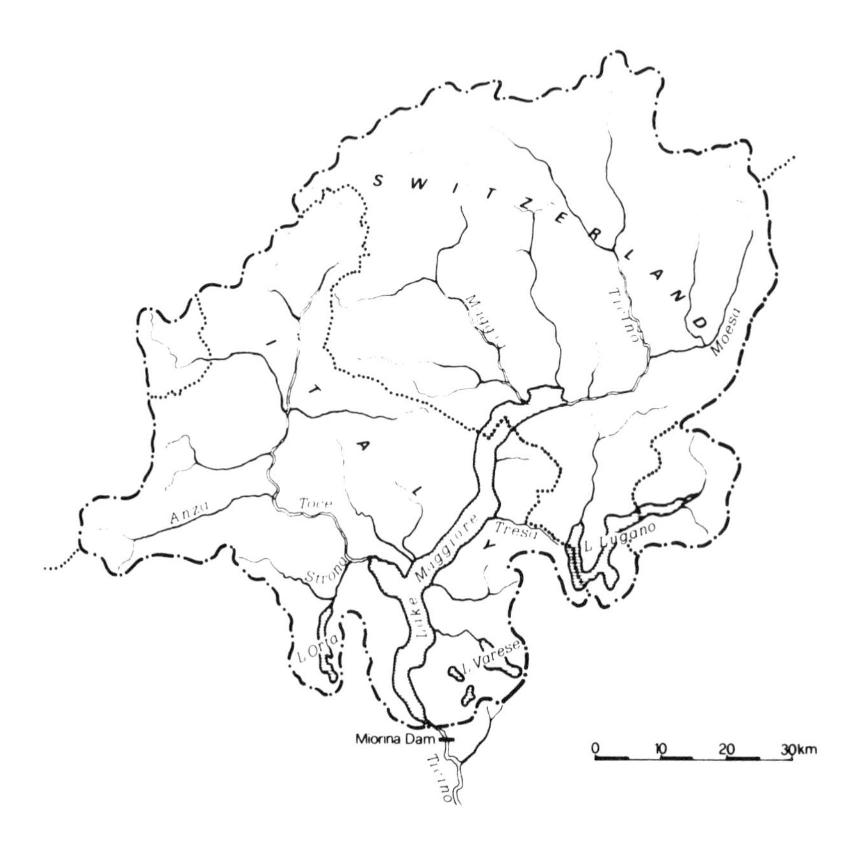

Fig. 4.9 — Lake Maggiore catchment.

surface, provided that the flows are within a given range, defined by the hydraulic characteristics of the irrigation network. Furthermore, East Sesia district has the option to take water from seven other unregulated sources (the main one of which is the Sesia river) which constitute about 65% of its overall supply.

Another main user is the Italian National Power Agency (ENEL), which utilizes water flowing out of Lake Maggiore for five power stations. Four of them are in-stream hydroelectric plants (for a total capacity of about 50 MW) located on different branches of the Ticino river, while the fifth is a large thermal station which requires at least 25 cubic metres of water per second for cooling. Given the importance of this plant, any policy which may result in a decrease of the flow at the station intake canal below that limit is to be considered unacceptable in practice.

The water remaining in the Ticino river after these diversions flows through a natural park, much visited and used by tourists.

To satisfy all these users, the storage in the lake would have to be relatively high during periods of water abundance, in order to have enough water on hand during the following dry seasons. Such a policy would

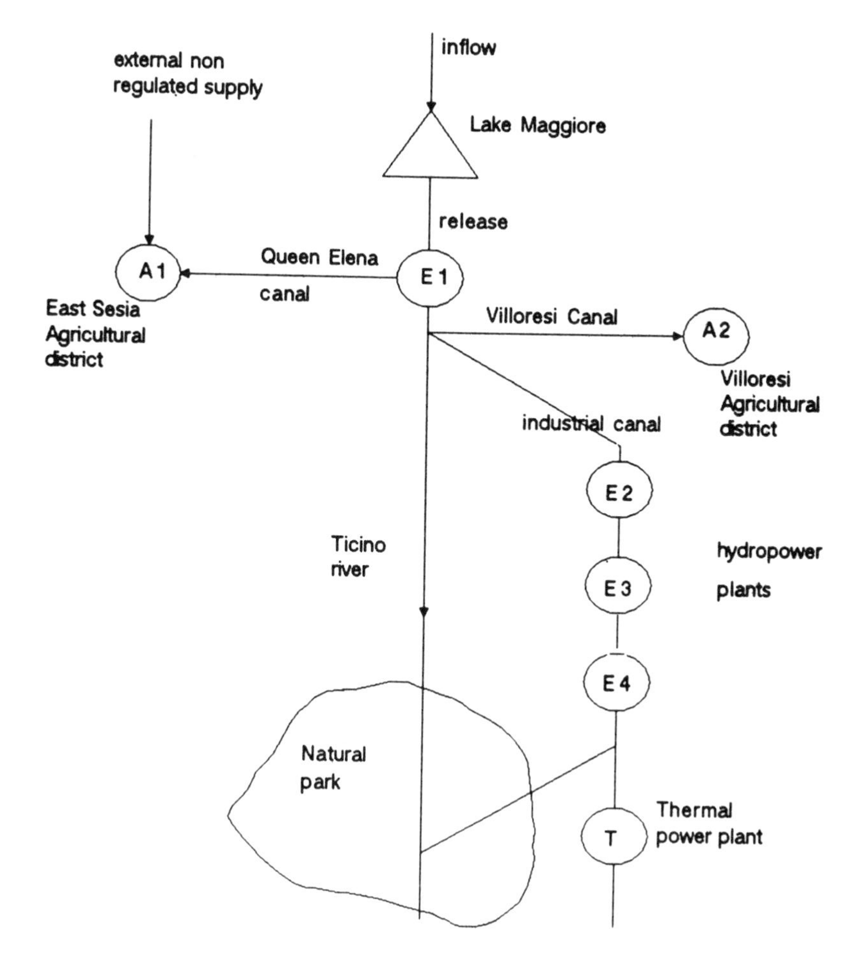

Fig. 4.10 — Main users of the Ticino river waters.

obviously conflict with the interests of the municipalities on the lake shores, which may suffer from dramatic flooding if the level of the lake increases too much. Tourism is the main activity in the 45 municipalities (15 of which are in Switzerland) surrounding the lake, but navigation is also an important support for their economic life.

Previous studies (Guariso and Laniado 1984, Bianchi and Laniado 1985) have estimated with a reasonable approximation the economic losses that would be suffered by all the downstream main users as a consequence of a deficit in the supply and those that would be incurred by upstream communities due to flooding. The mean annual value of all these damages over the past twenty years has been more than 50 billion Italian lire (at 1980 prices).

A number of other activities and social groups are also affected by the situation in the catchment (Guariso *et al.* 1986). Among these, the most important is tourism, both on the lake shores and along the upper part of the

Ticino River, which flows across the natural park, and wildlife conservation. These facets of the problem could not be quantified in monetary terms and have thus been measured by a physical indicator. For instance, downstream wildlife conservation has been measured by the number of days in which the oxygen concentration in the worst point of the Ticino would have been below a limit allowing normal trout fertility. The 11 indicators finally obtained are listed in Table 6.1.

Table 6.1 — Objectives and respective indicators of Lake Maggiore/Ticino system management

Sector	Indicator
Upstream Impacts	
Flood damages on the lake	Monetary damages due to flooding
Lake navigation	Nr. of days with lake level less than −30 cm
Sanitary problems	Nr. of days with lake level less than −30 cm
Swimming and recreation	Nr. of summer days with level greater than 60 cm or less than −30 cm
Damages to shore structures	Standard deviation of lake level below 0 cm
Wildlife conservation	Correlation with the natural lake regime
Downstream Impacts	
Agriculture	Monetary damages due to shortages
Power production	Power lost with respect to full plant capacity
Ticino flood damages	Mean peak flow
Sport fishing	Mean nr. of days with less than 4.5 mg/l of oxygen in the Ticino river
Swimming and recreation	Mean nr. of summer days with less than 50 m^3/sec in the river

When licensing the dam construction at the end of World War II, the Italian Ministry of Public Works, in agreement with Switzerland, set some limits to regulation in order to protect the interests of riparian municipalities. More precisely, the manager of the dam was left free to decide on the daily release from the lake, provided that the level of the lake, at the beginning of the day, fell within certain limits, i.e. the so-called control range. The Italian–Swiss commission, to which the definition of this range was finally committed, fixed the lower limit at a level which would prevent sanitary problems and guarantee navigation, and the upper limit at 150 cm during the winter period (November through March) and at 100 cm during the rest of the year. If the lake reaches the lower limit or exceeds the upper one, the manager has to take corrective action.

Furthermore, the management board, which is actually a committee formed by representatives of the downstream users, has been left completely free to decide upon the allocation of the water downstream from the lake. Since all the major abstraction canals along the Ticino are regulated, each supply may be fully controlled. This feature is of particular importance during droughts, since a portion of the standard hydroelectric supply is usually diverted to agricultural users. Conversely, in the case of abundant rain in the agricultural districts, more water is left for hydroelectric production or flows in the Ticino.

By setting the rules mentioned above and in particular the limits of the control range, the International Commission implicitly assumed a trade-off between all downstream and upstream activities and, in particular, between production and environmental conservation. Today, several components of this trade-off have changed: tourist activities on the lake have expanded; the natural park has been created; the construction of hydroelectric reservoirs in the catchment highlands has modified the inflow pattern; and more information has been gathered on the system dynamics. This is why several parties have suggested the revision of the old legal constraints, and various proposals have been put forward. VERBASIM is thus aimed at assisting the International Commission and the lake managers in replanning the Lake Maggiore–Ticino river catchment.

The design of VERBASIM

The system is oriented towards a restricted and well-defined set of users, who are practically all senior hydraulics engineers with a deep knowledge of the catchment characteristics and problems and little or no experience in using computers. It has been implemented on a VAX 750 computer using FORTRAN 77 and a medium resolution colour graphic terminal. Because the users are not computer experts, it has been designed in such a way that the use of the keyboard is minimized and the great majority of the input can be performed by cursor movements, with either a mouse, a graphic tablet, a light pen or a touch screen. On the other hand, all the problems of the catchment are described in a rather technical and quantitative way, since hydraulics engineers are familiar with operating rules, parameter values, and statistics.

The major decisions that the Commission may take concern:

— structural modifications of the system (such as varying the capacities of the canals), which normally require consistent investment;
— changes in the lake operation or the downstream allocation rules, which may require the installation of some telemetering stations and thus moderate expense, but do not necessitate modifications of the law;
— modifications of the legal framework of the lake operation, which require a new agreement between Italy and Switzerland (possibly with the agency managing the upstream reservoirs), but are completely costless.

Decisions which are likely to be adopted are normally a combination of the three types of changes described above, and thus it is essential to help the Commission understand and quantify the effects of each choice on each activity in the catchment. For this purpose, the program performs a simulation of the physical system based on an experiment selected by the user and defined by a sequence of inflows and by a set of changes of the three types cited above. The results of each simulation may in turn be examined under a number of different points of view, corresponding to different user-selectable outputs.

The modelbase thus contains a number of models, which describe the dynamics of the reservoir, the operating rules, the downstream water distribution system, the hydropower plants, the agricultural districts, the water quality of the downstream river and so on. While defining the experiment, the user has the possibility of setting the parameters defining these models, which represent the choices he or she wants to analyse, but he or she cannot change the way models are connected nor define new ones. In other words, only options on which the Commission and the lake managers have the power to take a decision have been left to the user. Other parameters, such as the crop mix in the agricultural areas, fall outside their responsibility and thus were considered as given. Given these restrictions, the access to the database is also highly simplified, since the only user choice is the period to simulate (and thus the sequence of inflows he or she wants to test); then all the data exchanges between the various models can be easily handled by the system.

The program has the classical computer-initiated user interface (see, for instance, Jensen 1983) and starts with a main menu which basically corresponds to the three logically independent subsystems of which it is composed: selection of the scenario, selection of the outputs and simulation. Pointing at one of these tasks, the user enters a second-level menu, which concerns the relative subsystem. Throughout the program, a valid selection is shown by the terminal through a variation of the colour of the item chosen.

When the first option is selected, the menu shown in Fig. 4.11 is displayed.

The choice of any of these items makes the program display a screen containing, on the upper part, a scheme explaining the meaning of the parameters to be changed. Below the scheme, more information on the effects of parameter changes is given in a short sentence, and, in the lower part of the screen, a table with the default values of the parameters (or the preceding ones in case a selection has already been made in the course of the same session) is presented (see Fig. 4.12). The user is requested to point to the parameter to be changed and then to input its desired value from the keyboard. Each time a new value has been digitized, the input is checked for correctness against a pre-specified range of allowable values. If the new value falls outside the range or does not comply with the required input, an appropriate error message is displayed.

In some cases, when a choice is made on the menu, an additional screen is displayed which details the meaning and possibilities of different options. For instance, a user who wants to test the effects of a new operating rule may choose from among four further possibilities shown on the screen. Each one represents a class of functions relating lake level and discharge. The parameters defining the specific rules within each class are entered in a subsequent frame, which explains how the form of the function is affected by the values of the parameters. Among the available choices are operating

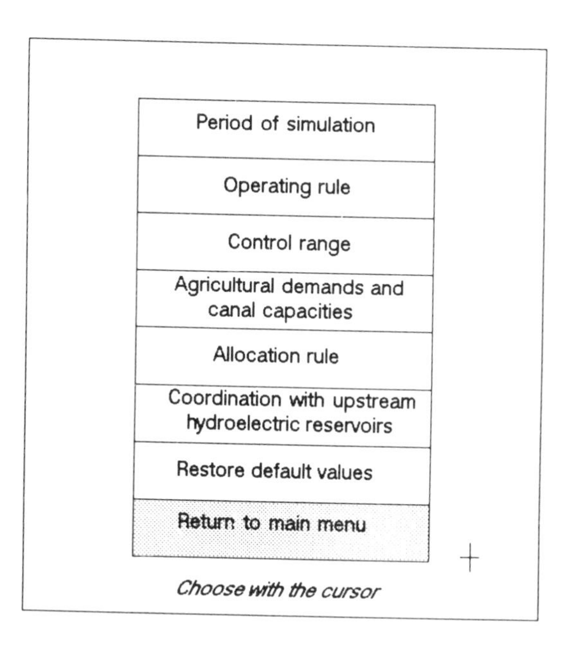

Fig. 4.11 — Menu of experiment definition in VERBASIM.

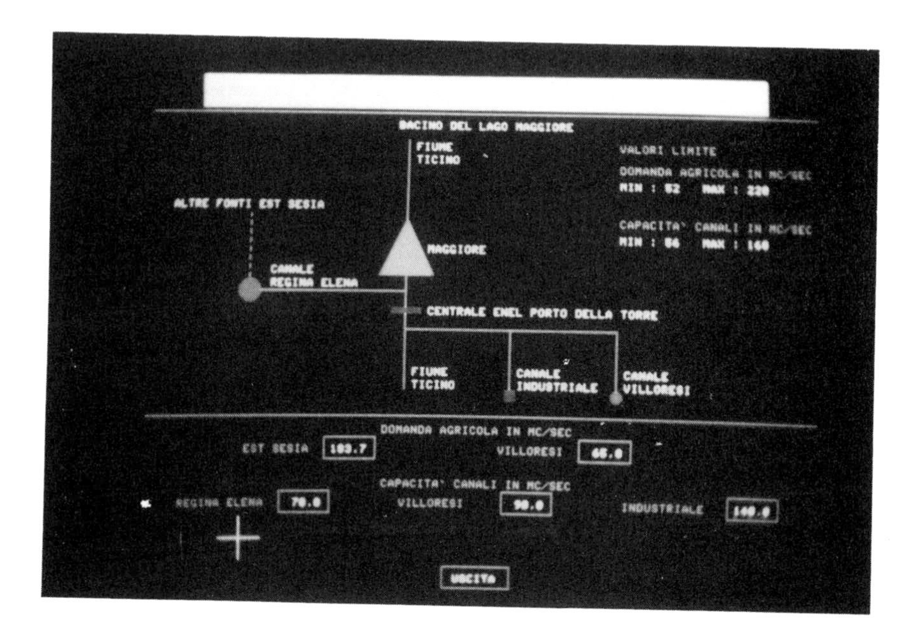

Fig. 4.12 — Definition of allocation rule in VERBASIM.

rules which favour only downstream or only upstream water users, or which constitute a compromise between the interests of the two groups (Guariso *et al.* 1985a). Other choices, which may be tested in relation to the lake discharge, are to dredge the river bottom at the lake outlet or to construct a canal which may bypass the dam, at least in very high floods, thus allowing more water to be released at the same lake level. Both these hypotheses have been strongly supported by Switzerland for a long time.

Four basic options are available for output corresponding to different degrees of detail.

The first type of output shows on one page the performance of the system from the point of view of the four main users: East Sesia and Villoresi agricultural districts, the National Power Agency and the communities on the lake shores. The yearly values of the damages incurred by each user for the entire simulation period is displayed, together with the mean values and the totals.

The second possibility is a duration curve of the daily flows (or levels) which represents the fraction of time in the simulation period in which the flow (level) has been greater than a certain value. It is a representation which is very familiar to hydraulics engineers, and thus helps them to understand the productivity of a certain flow or the danger of a certain level.

The third form of output is a graph of the daily values of flows or levels, which allows a detailed analysis to be made of management performance even during a particularly critical episode. Although the expected values of performance indicators are usually considered to be the main objective of a water management system, the decision-maker is often more sensitive to extreme situations which may cause dramatic damages (Guariso *et al.* 1984). As pointed out above, an intuitive and simple criterion to judge a certain proposal is to evaluate its performance during a particular flood or drought that has been experienced in the past.

The final type of output is a classical numerical one, particularly for all the indicators that have only a statistical meaning, such as those representing the impacts on the minor activities taking place in the basin.

The output may be stored for further comparison with that generated in another experiment (see Fig. 4.13). In this way, one can easily search for new proposals which improve the present situation, at least from a certain point of view. Furthermore, all of the outputs may always be compared with two reference conditions: the historical behaviour of the system in the simulation period; and the so-called natural conditions, i.e. those existing before the construction of the dam.

Outcome of the experience

VERBASIM has been extensively tested and experimentally used in continuing education courses for engineers, and it has been recently made officially available to the International Commission for showing the consequences of any proposal which will be raised during their discussions. From this first intensive period of experimentation, some preliminary conclusions can already be drawn.

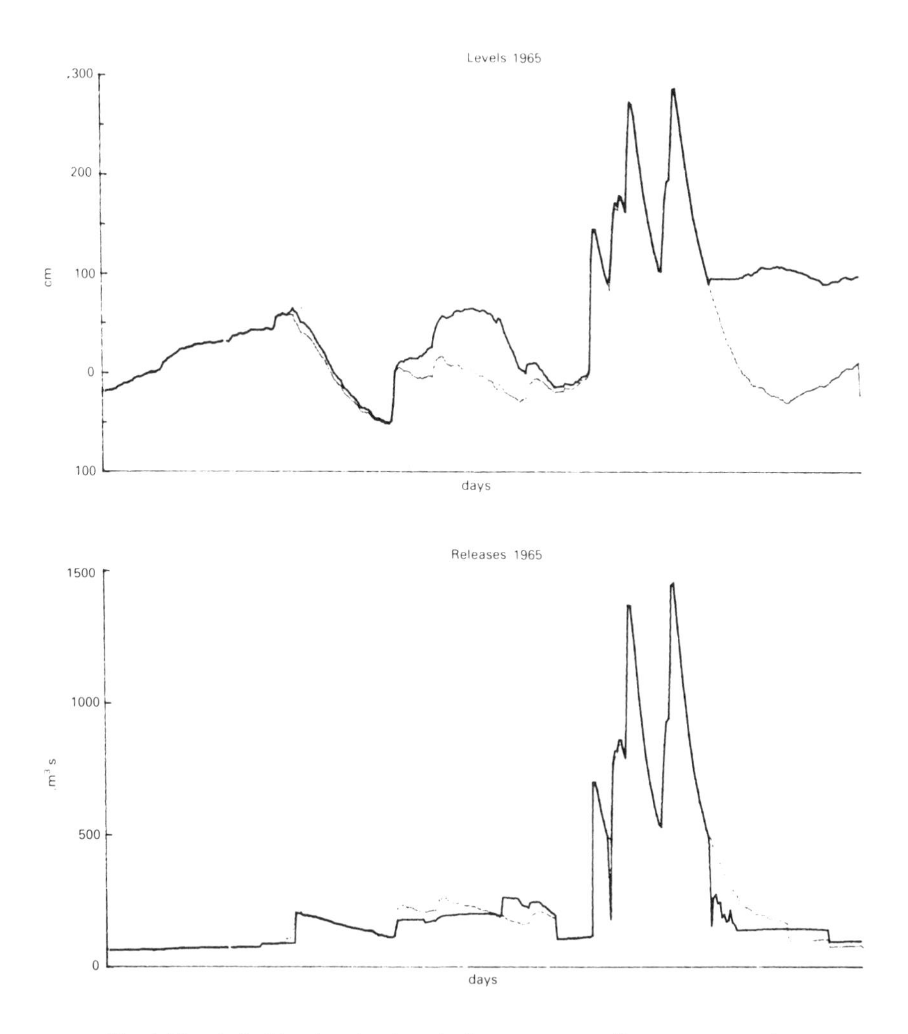

Fig. 4.13 — Lake Maggiore levels and releases corresponding to two scenarios.

It has been shown that various physical and legal constraints on the system may be profitably revised. For instance, there is a clear imbalance between the capacities of the main irrigation canals and the water requirements of the respective agricultural users. The two canals were in fact designed when the crop mix was very different from the present one. A 30 cm decrease of the upper limit of the control range in October and November may decrease mean flood damages by about 12%, practically without affecting downstream uses. Though being definitely more expensive, the construction of a bypass canal which can be opened at high lake levels appears to be the most effective decision. By increasing the discharge to about 100 cubic metres per second outside the control range, it is possible to reduce mean flood damage to one third of the value that has been sustained over the last 20 years.

However, the most striking result obtained using VERBASIM is the determination of a new control range and a new allocation rule that may improve (or at least not damage) the situation of all users, i.e. a solution which is more efficient than all those put forward up to now and thus must be acceptable to all parties. This clearly indicates that such programs may sometimes help promote an open-minded discussion and more cooperation, which can lead to solutions that improve everybody's conditions. Obviously, a final choice will also be based on other political considerations which are outside the physical system simulated on the computer. However, these decisions will certainly benefit from the systematic analysis of all the implications and effects, some of which are certainly not intuitive and not easily solvable by analytical methods.

The main difficulty encountered in the design of the system was to find an acceptable trade-off between answering any possible question concerning lake and river behaviour and achieving a speed sufficient to allow the user to repeat the simulation several times without being bored by long response times. Strict cooperation with the final system users is probably the essential prerequisite in achieving this result (Maguire 1985). The system implemented for Lake Maggiore seems to have been accepted by the users for whom it has been designed, but it must be pointed out that it is the result of almost ten years of study and cooperation.

4.6.3 APC: controlling air pollution using forecasts

In Italy, as in many other countries, legislation sets precise limits as to the quality of the air in the neighbourhood of any large pollution source, such as a thermal power plant. These limits are usually translated into standard concentrations of different pollutants, which can only be topped with a given frequency. One such limit is that the daily average of the ground level concentration around a plant cannot exceed a given value (to be precise, 250 $\mu g/m^3$ for SO_2, the pollutant discussed in the following text) more than 7 days a year.

Obviously, such concentrations are not only the result of a certain level of electricity production and thus of a certain pollution emission, but are also due to particular meteorological conditions, which produce a so-called fumigation episode. To summarize the complex dynamics of this problem in a few words, when temperature conditions are such that there is an inversion in the derivative of temperature with respect to elevation (which may in turn be due to a particular pattern of solar radiation and a stable situation of the atmosphere), the plume emitted by the stack is captured close to the plant, and the resulting ground-level pollution concentration in the surrounding area is considerably higher than in normal non-inversion situations. Depending upon the site and the local meteorology, these conditions may be more or less frequent, but it is clear that they are the major concern of the managers responsible for satisfying air-quality standards.

There are several means to avoid exceeding the limits, even in adverse conditions. One may filter the pollutant, use a less polluting fuel, reduce the production at that site and so on. Each of these measures has been widely used and proved to be effective, but such measures are very expensive. One additional possibility exists for plants that can work with a different fuel mix, namely with fuels with different sulphur contents. In this case, when the risk of exceeding the standard is considerable, the manager may decide to switch to the cleaner fuel in order to reduce emissions, without lowering the production of the plant. The cleaner fuel obviously costs more than the standard one, and thus the number of substitutions and their amount must be minimized.

To help the managers of a thermal power plant in deciding when to take such a step and on the level of fuel substitution a software package (APC) was developed by ENEL, the National Italian Power Agency (Brusasca *et al.* 1983, Pagliari *et al.* 1984) on the basis of a long series of studies (Melli *et al.* 1981, Bacci *et al.* 1981, Bolzern and Fronza 1982, Brusasca and Finzi 1986).

APC modelbase

The structure of APC is described in Fig. 4.14. A computer collects meteorological and pollution data from a telemetering network, then a model is used to predict the meteorological conditions over the next few hours, and another model uses these forecasts for computing the expected values of pollutant concentration for the same interval of time. In case these values exceed a given limit p, which is lower than the legislative standard to allow for forecasting errors, the system suggests a certain fuel substitution with the criterion of minimizing fuel costs.

The features of the various models used in APC are summarized below; interested readers will find more details in the referenced literature.

The model for pollution forecast is a very simple autoregressive model with exogenous input (ARX) of the type

$$\hat{p}(t + Dt) = \alpha(t)\, p(t) + \beta(t)\, u_s(t + \mathrm{D}t)$$

where $\hat{p}(t + \mathrm{D}t)$ is the pollution forecast for time $t + \mathrm{D}t$, $p(t)$ is the pollution measured at time t (when the forecast is issued), $u_s(\mathrm{t} + \mathrm{D}t)$ is the emission scheduled for time $t + \mathrm{D}t$ to produce the requested power with the standard fuel, $\alpha(t)$ and $\beta(t)$ are time-varying parameters which in turn depend on the atmospheric stability and wind direction class forecast.

The model used to predict the stability class is in turn based on the wind speed at time t, the total radiation since sunrise and the status of the air at the end of the previous night, while wind direction is assumed to be the one maximizing the probability of occurrence conditioned to recent and current wind records (see again Bacci *et al.* 1981, for details).

As in the case of WODA, one may simplify the class of models to use by ignoring the dependence of certain parameters on certain atmospheric variables and obtain in this way a much simpler predictor. For instance, assuming Dt is half an hour and $\alpha(t)$ does not depend upon any other variable, one obtains a cyclostationary autoregressive predictor, which has been extensively used in actual applications.

In practice, the situation is more complex, since fuel substitution takes a

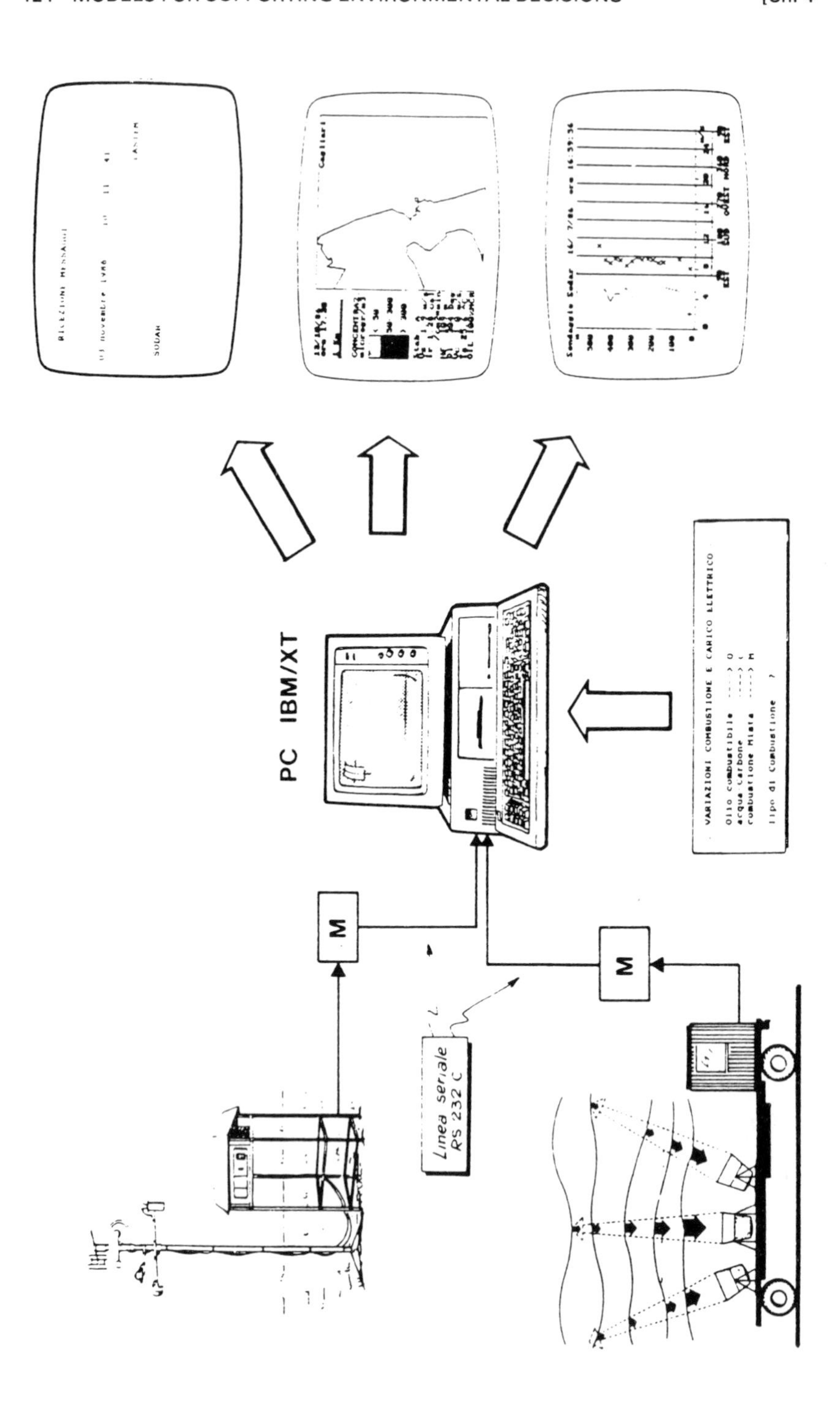

Fig. 4.14 — APC pollution forecast and control system.

certain amount of time to take effect, and thus the forecast must be made for a couple of intervals ahead. This may be done by assuming the first prediction as new datum and implementing all the models again from this new condition.

The criterion for fuel substitution is the minimization of fuel costs. Thus, assuming q_h is the quantity of standard fuel with a high sulphur content and q_l is the quantity of cleaner fuel with a low sulphur content, r_h and r_l are their respective SO_2 emission rates and s_h and s_l are their prices, the following optimization problem is solved at each time step Dt:

$$\min_{\{q_h,\, q_l\}} \quad (s_h\, q_h + s_l\, q_l)$$

subject to

$$\hat{p}\,(q_h, q_l) = \alpha\, p + \beta\,(r_h\, q_h + r_l\, q_l) \leqslant \underline{p}$$

$$q_h + q_l = u_s / r_h$$

$$q_h, q_l \geqslant 0$$

The first constraint indicates that the limit value $\underline{p}$ must not be exceeded, and the second that the production must follow the schedule (the scheduled quantity of fuel is burned).

The solution to this problem is very simple and constitutes the so-called linear roll-back policy, i.e., the quantity of standard fuel which is replaced by the cleaner one is proportional to the forecast violation of the standard $(\hat{p}(u_s) - \underline{p})$ through the coefficient β. This means that the exact limit would be reached if the forecast is perfect.

In view of the inaccuracy of forecasts, however, the limit $\underline{p}$ cannot be set at a value equal to that allowed by the law. The choice of the value of $\underline{p}$ is thus, in principle, a planning problem which has to be solved before implementing the real-time control procedure and implies the determination of the trade-off between the risk of pollution and the cost of making unnecessary substitutions. However, if a fixed value for $\underline{p}$ could be set, then the overall procedure can be implemented without human intervention. This is in practice impossible, since the quality of the forecast varies widely under different meteorological conditions; thus APC only suggests a range of values for q_l (cleaner fuel), but the plant manager has responsibility for the final choice.

To assist the manager in making that choice, some short-term deterministic models have also been implemented. They allow a user to compute the stability of the atmosphere, or the plume rise for each stack according to Briggs' formulas, or empirical formulas to check if the plume may exceed a detected thermal inversion, or to use a Gaussian plume model to compute ground-level SO_2 concentration.

The user interface

Two versions of the APC have been implemented, one running on a minicomputer (Brusasca *et al.* 1986) and the other on a Personal Computer. The differences are due mainly to the different meteorological networks connected with the two computers. Both versions use a menu-driven interface; the user can select the models to run but not the way they are connected, and he or she cannot create new models. On the other hand, APC has a comprehensive data storage and retrieval system so that, before taking the final decision, the manager can also examine the individual data which have led to the suggested solution and, more important, how similar episodes have developed in the past. Although, in fact, mathematical models are exactly based on these past episodes, usually they represent a mean behaviour, while the manager is often able to go back to a single specific episode which appears very similar to the actual one. The standard output is represented by graphs of vertical profiles of wind and temperature, a plot of the concentration at each ground sensor over the last 24 hours, and maps of the plume dispersion.

The interface module is not very sophisticated since the target users are senior engineers with a certain amount of experience with computers and a deep knowledge of the problem at hand. Only a few people have thus used APC up to now, and it has not been employed for training purposes.

Outcome of the experience

The APC minicomputer version has been in use since 1984 at the Turbigo power plant, while the Personal Computer version was installed in 1986 at S. Gilla and will soon be used also at Ostiglia. Turbigo and Ostiglia are located in the Po plain in Northern Italy and are close to some small country towns. At both sites, the mean wind speed is very low, and calm conditions occur 17% of the time in Turbigo and 5% in Ostiglia. S. Gilla is in Sardinia and has a more favourable atmospheric situation.

Turbigo has a full capacity of 1365 MW with six sections. Three generate 320 MW and are connected to 150 m high stacks. The other three sections have 260, 75, and 70 MW capacity and stacks of 96 and 48 m. S. Gilla is a small experimental plant located near Cagliari and has two sections each of 35 MW connected to a 35 m stack. Ostiglia has four 320 MW groups (and thus a total capacity of 1280 MW) connected to stacks 200, 170, and 120 m high.

All plants normally burn oil fuel with a 3% sulphur content, but may shift in less than one hour to a cleaner fuel with 1% sulphur. They have a monitoring network composed of five meteorological and pollution stations, situated from 0.5 to 5.5 km from the stacks, mainly in the direction of prevailing winds. Furthermore, they are equipped with remote sensing instruments for meteorological conditions: a Doppler SODAR and a Radio Acoustic Sounding System (RASS).

The SODAR uses three antennas and measures the backscattering echo from the atmosphere once a powerful acoustic pulse is emitted upwards. Thus, using sophisticated software, it is possible to determine the wind speed from 50 m to about 1000 m elevation with a precision of about 0.3 m/s and the wind direction with a precision of 3° with a spatial resolution of 50 m. RASS tracks the speed of a short acoustic pulse and determines, in this way, the temperature profile from 50 m up to 1000 m with a resolution of 20 m. All the instruments are connected by serial RS 232 lines and are polled by the computer every half hour. A serial line is also connected to the plant's main control to acquire data on load, smoke temperature, SO_2 emission and other information.

APC has been installed at both sites, and users have been trained in the past few years. The managers seem to like the system, so that ENEL is planning to extend it to other plants. However, a fair evaluation of the experience is impossible since production, particularly at Turbigo, has been considerably below full capacity in the last few years, so that peak pollution episodes have almost disappeared. Apparently it has been cheaper for ENEL to buy power from abroad than to produce it in these plants. However, this situation seems to be coming to an end (particularly after the Italian Government's decision to close all existing nuclear plants and halt plans for new ones) and thus more complete records of real decisions and those suggested by APC will be available for comparison.

4.6.4 SILVIA: A decision support for EIA

As already mentioned in this chapter, Environmental Impact Assessment (EIA) is a complex and interdisciplinary task aimed at pointing out all the effects of any project or plan in such a way that a decision maker may better consider all facets of the problem and improve the transparency and the participation of all the social groups involved (see Lee 1983).

SILVIA (Colorni and Laniado 1986), an Italian acronym for 'Interactive Software for Environmental Impact Assessment', is a large and long-lasting effort sponsored by the Italian Ministry of Environment for developing software to assist the various phases of EIA. More precisely, three phases have been identified as suitable areas for software development (they are closely related to the decision phases defined in Chapter 1): a preparatory phase, an analysis phase, and a decision phase. The preparatory phase involves gathering information on the type, size and location of the project or plan to be analysed, checking what the laws in force state about the particular case; checking how the problem has been tackled on similar occasions in the past; ascertaining which data on the environment are pertinent to the case under discussion and where they are available; and finding out which social groups and environmental sectors are involved.

The second phase consists of defining available alternative decisions; the elementary actions implied by each of them; all environmental sectors which may be directly or indirectly impacted by each action; and a suitable set of indicators for each one of these sectors. Finally, models (packages) or experience is used to evaluate in quantitative or qualitative terms the effect of each action on each indicator.

The decision phase involves the definition of proper criteria to eliminate some of the alternative decisions in order to arrive at one or at least very few projects from which the final (political) choice has to be made.

The structure of SILVIA

SILVIA is composed by three different packages, each covering one of the phases mentioned above. The first is an expert system (Belotti 1987) which acts as an intelligent front end to a database containing the legislation on EIA, records of previous analyses, and catalogues of available data and models. It helps the user in understanding the bureaucratic procedures required to carry out the assessment correctly, what has been done in the past in similar cases, and where and how he/she can get quantitative or qualitative estimates of the impacts.

The second phase is supported by a package which allows alternative projects and environmental sectors and indicators to be defined, and then, using a series of coaxial matrices (Canter 1977), results in the definition of a single final matrix, the impact matrix, which has a column for each alternative and a row for each environmental indicator. The program starts from a set of matrices (one for each alternative project) which have a column for each elementary action implied by the alternative and a row for each environmental sector; and the number or columns or rows expands or shrinks according to the user's choices (i.e. to specify certain points better or to aggregate certain actions or sectors).

Each element of the intermediate matrices may contain qualitative or quantitative evaluations, while a conversion table (defined by the user) allows the qualitative terms to be translated into quantitative estimates in the impact matrix.

The aim of the third and last package is to rank the alternatives using multicriteria techniques on the impact matrix (see Figure 4.16, p. 132).

The decision models

The first operation performed by the latter package on the impact matrix is the elimination of the non-efficient alternatives according to Pareto. This operation is objective and may be automatically performed by the program.

The second step is normalization, i.e. the transformation of all the impact values into a similar range of values (normally from 0 to 1). This is necessary since the numerical differences between environmental indicators may be of several orders of magnitude. The normalization must be driven by the user, who has to decide, for each environmental indicator, how to scale values to the unit interval. The program allows him or her the choice between a standard method (linear scaling), piece-wise linear and user-defined non linear functions. This very preliminary step already introduces subjective judgement in the evaluation and may be heavily distorted by the presence of very extreme alternatives. Their indicators must obviously be transformed into the value 1, which may reduce the differences between the other alternatives to very small decimals.

The third step is the determination of weights to allow an aggregate evaluation of the performance of each alternative. In other words, an index I_k is associated with each alternative k computed as

$$I_k = w_1 i_{k1} + w_2 i_{k2} + \ldots . + w_n i_{kn}$$

where w_j is the weight of the environmental indicator j and i_{kj} is the impact of the kth alternative in that sector and

$$\sum_{j=1}^{n} w_j = 1$$

The weights w_j may be directly chosen by the user or automatically computed by the program, either using the entropic method suggested by Hwang and Yoon (1982) or approximating, in a consistent way, a matrix or relative weights of each pair of indicators supplied by the user. As already noted in Chapter 2, while it is very difficult for a user to assign weights when the number of indicators is high, it is usually feasible to ask him or her for the relative importance of two different indicators. In particular, if w_{ij} is the relative importance of indicator i with respect to j, the properties of reciprocity ($w_{ij} = 1/w_{ji}$) and of consistency ($w_{ij} = w_{ih} w_{hj}$) are unlikely to be satisfied. However, one may obtain a similar, but consistent, weighting, for instance by finding the values which better approximate those supplied by the user according to a least-square's criterion. Other possibilities, supplied by the program, are the average, overlapping and eigenvalue methods (Hwang and Yoon 1982).

The index I_k already provides a ranking of the alternatives, but only a sensitivity analysis of the values of the weights provides information on the reliability of this ranking. The program can in fact compute the interval of values of each weight for which the ranking does not change. This information gives substantial advice on the robustness of a choice and is particularly useful for all those weights which have been set in a very heuristic manner. If a small shift in the weight values alters the ordering, a more refined evaluation of the importance of that particular indicator is necessary. If, however, the value of a weight has to vary by one or more orders of magnitude in order to induce a change in the best alternative, there is no need to further examine the weighting of that particular indicator.

Another selection may be suggested by the program using the ELECTRE method (Goicoechea *et al.* 1982). For this purpose a concordance index c_{kl} and a discordance index d_{kl} may be computed between each pair of alternatives k and l. The first is a measure of the satisfaction of choosing k with respect to l and the second of the regret of making that choice. They can be computed as

$$c_{kl} = \Sigma_j w_j \quad \text{for each } j \text{ such that } i_{kj} \text{ is bettter than } i_{lj}$$

$$d_{kl} = \frac{\max_j w_j\,(i_{kj} - i_{lj})}{\max_J w_J\,(i_{kJ} - i_{lJ})} \qquad \begin{array}{l}\text{for each } j \text{ such that } i_{lj} \text{ is better than } i_{kj} \\ J = 1,2,\ldots,n\end{array}$$

The user may then fix two thresholds S_c and S_d for concordance and discordance indexes and eliminate all the alternatives l such that there exists an alternative k with

$$c_{kl} \geqslant S_c$$

$$d_{kl} \leqslant S_d$$

This condition is sometimes referred to as 'weak dominance' of k over l.

Once more, it is often more interesting to analyse the sensitivity of the choice to the values of the thresholds and thus present on the screen in the plane S_c, S_d those values for which an alternative will be weakly dominated by some other one. The larger the portion of the plane for which this happens, the less regrettable will be its elimination.

Finally the program allows the calculation of three more indices to order the alternatives. An overall concordance index is computed as

$$C_k = \Sigma_j\, c_{kj} - \Sigma_i\, c_{ik}$$

and measures the satisfaction of choosing alternative k. An overall discordance index

$$D_k = \Sigma_j\, d_{kj} - \Sigma_i\, d_{ik}$$

represents the total regret in choosing k (the lower D_k is the more appealing is k); and finally a weighted sum of the two

$$E_k = \alpha\, C_k + (1 - \alpha)\, D_k$$

may provide a unique composite index to rank the alternatives.

The user interface

The dialogue between user and program takes place in SILVIA through menus, matrices and graphics (Colorni and Laniado, 1987a). The menus show the possible options, and sometimes graphics immediately portray the effects of a certain choice. For instance, during the normalization step, each time the user selects or defines a function it is immediately shown on the screen. Also some of the results can be graphically displayed in order to show the dominance or weak dominance between various alternatives, for instance.

An extensive help facility has been provided in order to make the user conscious of the effects of each individual choice during the process. Each time there is an automatic option to perform a certain step, there are warnings indicating the drawbacks of the method implemented.

A key feature of the interface of the decision module is that it has been designed in order to be as objective as possible and to discourage the simplistic approach of setting some weights and then picking up the first alternative in the ranking. First of all, the process is aimed at pointing out ineffective decisions instead of stressing the better ones. Second, great relevance is given to the robustness of the ranking, i.e., to sensitivity analyses, and not to absolute values. Third, the alternative with the higher ranking is never referred to as the 'best' or the 'most effective' or anything else which may induce the user to stop the analysis at that point.

The interesting aspect of this package lies indeed in the possibility of repeating the analysis with a number of different indicators, alternatives, and weights in order to perceive in a better way (and make evident to interested parties) all the critical points of the decision process.

Outcome of the experience

SILVIA, and more precisely the packages for the second and third phase, has been implemented on a PC mainly using Fortran and Assembler since 1986 and has been experimentally used in a number of environmental impact assessments and for developing a gaming session during environmental impact courses.

One of the major cases in which it was used was a project for a freeway around the city of Mantua (Mantova) in Northern Italy (Colorni and Laniado 1987b). The problem was the choice between four alternative routes (see Fig. 4.15), and the environmental sectors originally chosen were

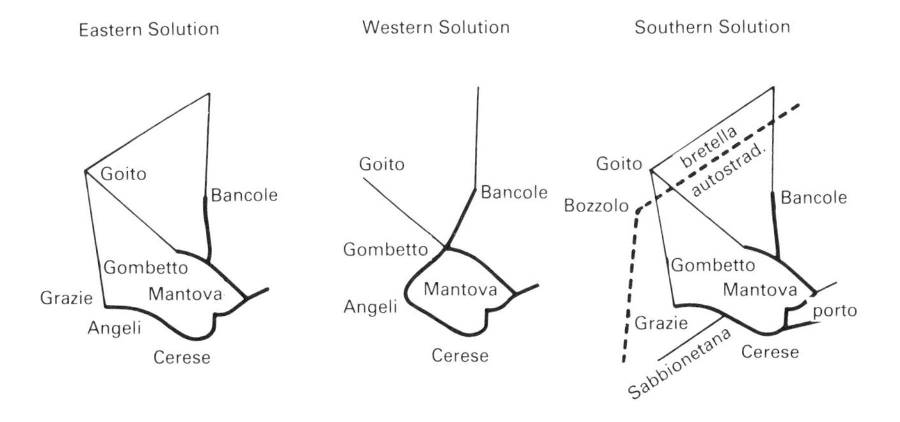

Fig. 4.15 — Three proposals for the freeway around Mantua (Mantova). The fourth project was a minor variation of the second one.

12 in each of the 39 elementary areas which the freeway was to cross. The initial matrix thus had 468 rows and 5 columns (the 'null' hypothesis, i.e. no freeway, was added for comparison). Through the analysis package, a final impact matrix of 12 rows and 5 columns was finally computed (Fig. 4.16).

Polytechnic of Milan, Department of Electronics — Project SILVIA

DATA INPUT AND RETRIEVAL

Sector	Null hypothesis	East	West	South	West variation
Land	0.50	0.20	0.15	0.25	0.15
Aquifer	0.50	0.35	0.30	0.35	0.30
Surf. water	0.50	0.40	0.25	0.40	0.40
Air	0.50	0.45	0.40	0.45	0.40
Vegetation	0.50	0.30	0.15	0.30	0.15
Fauna	0.50	0.45	0.45	0.45	0.45
Recreation	0.50	0.45	0.45	0.45	0.45
Landscape	0.50	0.30	0.10	0.10	0.15
Agriculture	0.50	0.35	0.35	0.35	0.35
Hist. sites	0.50	0.15	0.15	0.15	0.15
Housing	0.50	0.25	0.25	0.25	0.25
Local econ.	0.50	0.50	0.50	0.50	0.50

Fig. 4.16 — The impact matrix of the Mantua freeway.

The decision package showed that one of the alternatives was preferable to the other for a very large spectrum of criteria used, but the actual choice of the municipality of Mantua was for a different solution.

This should not come as a surprise since, as already stated, SILVIA aims at showing all of the facets of a decision in a systematic way and pointing out explicitly the trade-offs between the various criteria and choices. It is easy to say that the number of choices the user has to make when using the package appears to be more complex than the original choice between alternative projects and that the final result obtained through the package has exactly the same degree of subjectivity.

These are not criticisms, since SILVIA is not intended to arrive at the 'most desirable' solution, but rather is expected to help the user make a decision in a more conscious manner and favour transparency, and thus participation, in the decision process. This is a common characteristic of EDSS, which, according to Geoffrion (1976), they share with many other applications of mathematical models and computer science: 'The purpose of mathematical programming is insight, not numbers'.

REFERENCES

Bacci, P., Bolzern, P. & Fronza, G. (1981) A stochastic predictor of air pollution based on short-term meteorological forecasts, *J. Applied Meteorology* **20**, 121–129.

Barnwell, T. O., Brown, L. C. & Marek, W. (1986) *Development of a prototype expert advisor for the enhanced stream water quality model QUAL2E*. Internal Report, US EPA, Athens, Georgia, USA.

Belotti, G. (1987) *An expert system for environmental impact assessment*. Master Thesis, Politecnico di Milano (in Italian).

Bianchi, A. & Laniado, E. (1985) *Estimation of flood damages on lake shores*. Report of the Institute of Hydraulics, Politecnico di Milano (in Italian).

Birta, L. G. (1984) Optimization in simulation studies. In: Ören, T. I., Zeigler, B. P. & Elzas, M. S. (eds), *Simulation and Model-Based Methodologies: An Integrative View*, NATO ASI Series Vol. 10. Springer-Verlag, Berlin, pp. 185–216.

Bolzern, P. & Fronza, G. (1982) Cost-effectiveness analysis of real-time control of SO_2 emission from a power plant. *J. Environmental Management* **14**, 253–363.

Box, G. E. P. & Jenkins, G. H. (1970) *Time Series Analysis Forecasting and Control*. Holden–Day, San Francisco.

Brusasca, G., Elisei, G., Maini, M. & Marzorati, A. (1983). Acoustic remote sensing for environmental control in thermal power plants. In: *Proc. II Int. Symp. on Acoustic Remote Sensing of the Atmosphere and Oceans*, XXI, 1–12, Rome.

Brusasca, G. & Finzi, G. (1986) Stochastic models for real-time SO_2 pollution forecast around thermal power plants. In: Zannetti, P. (ed.), *Proc. ENVIROSOFT 86*. Computational Mechanics Publ., Southampton, UK, pp. 177–191.

Brusasca, G., Marzorati, A. & Tinarelli, G. (1986). Software for real-time acoustic remote sensing instrumentation and atmospheric pollution control. In: Zannetti, P. (ed.), *Proc. ENVIROSOFT 86*. Computational Mechanics Publ., Southampton, UK, pp. 759–769.

Canter, L. W. (1977) *Environmental Impact Assessment*. McGraw–Hill, New York.

Cohon, J. L. (1978) *Multiobjective Programming and Planning*, Academic Press, New York.

Cohon, J. L. & Marks, D. H. (1975) A review and evaluation of multiobjective programming techniques. *Water Resources Research* **11**, 208–219.

Colorni, A. & Laniado, E. (1986) A software package for Environmental Impact Assessment. In: Zannetti, P. (ed.), *Proc. ENVIROSOFT 86*. Computational Mechanics Publ., Southampton, UK, pp. 745–757.

Colorni, A. & Laniado, E. (1987a) A decision support system for choosing among alternative projects. In: *Proc. of the Seminar on Multipurpose Agriculture and Forestry*. Wissenschaftverlag Vauk, Kiel, pp. 39–51.

Colorni, A. & Laniado, E. (1987b) Application of a decision support system to the environmental impact assessment of the 'Tangenziale di Mantova' (in Italian). In: *Proc. Italian Conf. of Regional Sciences*, Cagliari.

De Buyser, R. & Spriet, J. A. (1988) OPTISIM: An optimization–simulation environment for model building. In: *Proc. European Simulation Multiconf.*, Nice, 1–3 June 1988. SCS, Ghent, Belgium, pp. 16–21.

De Simoni, G., Fanello, R. & Guariso, G. (1986) Computer aided planning of a multipurpose river basin. In: Zannetti, P. (ed.), *Proc. ENVIROSOFT 86*. Computational Mechanics Publ., Southampton, UK, pp. 421–434.

Dobbins, W. E. (1964) BOD and oxygen relationships in streams, *J. Sanit. Eng. Div. ASCE* **90**, 53–78.

Elzas, M. S. (1984). System paradigms as reality mappings. In: Ören, T. I., Zeigler, B. P. & Elzas, M. S. (eds), *Simulation and Model-Based Methodologies: An Integrative View*. NATO ASI Series Vol. 10. Springer–Verlag, Berlin, pp. 41–68.

Fedra, K. (1985) *Advanced decision oriented software for the management of hazardous substances* (Parts I–III, VI), CP-85-18/85-50/86-10/14, International Institute for Applied Systems Analysis (IIASA), A-2361, Laxenburg, Austria.

Fronza, G. & Garofalo, F. (1979) Decision models for planning and management of water resources (in Italian). In: *Proc. Seminar on River Basin Planning*. Univ. of Genoa.

Gandolfi, C., Guariso, G. & Rinaldi, S. (1988) Multicriteria reservoir control: Experience on an Italian lake. In: *Proc. IFAC Symp. on Systems Analysis Applied to Management of Water Resources*, pp. 153–160.

Gass, S. I. (1979) Computer Model Documentation: A Review and an Approach. *NBS Publ. 500-39*, US Dept of Commerce.

Geoffrion, A. M. (1976) The purpose of mathematical programming is insight, not numbers. *Interfaces* **7**, 1.

Goicoechea, A., Hansen, D. R. & Duckstein, L. (1982) *Multiobjective Decision Analysis with Engineering and Business Applications*. John Wiley, New York.

Golden, B. L., Rothschild, B. & Assad, A. A. (1987). A microcomputer-based decision support system for multi-species fishery management. In: Lev, B. (ed.), *Strategic Planning in Energy and Natural Resources*. Elsevier Science Publ., Amsterdam, pp. 333–339.

Guariso, G., Gandolfi, C. & Pirovano, G. (1988) A PC environment to support groundwater managers. In: Zannetti, P. (ed.), *Proc. ENVIROSOFT 88*. Computational Mechanics Publ., Southampton, UK, pp. 189–198.

Guariso, G. & Laniado, E. (1984) Economic consequences of agricultural deficits: the case of Ticino (in Italian). *Rivista di Economia Agricola* **39**, 659–676.

Guariso, G. & Werthner, H. (1986) A computerized inventory of water resources models, *Environmental Software* **1**, 40–46.

Guariso, G., Orlovski, S. & Rinaldi S. (1984) A risk-averse approach for reservoir management. In: *Proc. IX IFAC World Congress*, Vol. IV, pp. 183–188, Budapest, Hungary.

Guariso, G., Laniado, E. & Rinaldi, S. (1985a) The management of Lake Maggiore: conflict analysis and the price of noncooperation. *OR Spektrum* **7**, 101–109.

Guariso, G., Rinaldi, S. & Soncini-Sessa, R. (1985b) A decision support system for the management of Lake Como. *European J. Operational Research* **21**, 295–306.

Guariso, G., Rinaldi, S. & Soncini-Sessa, R. (1986) The management of Lake Como: A multiobjective approach, *Water Resources Research* **22**, No. 2, 109–120.

Hall, W. J., Haimes, Y. Y. & Freedman, H. (1975) *Multiobjective Optimization in Water Resources Systems*. Elsevier, Amsterdam.

Halpern, P., Roberts, S. M. & Lopez, L. (1987) An incidence-matrix-driven panel system for the IBM PC, *IBM Systems J.* **26**, 201–214.

Herfindhal, O. C. & Kneese, A. V. (1974). *Economic Theory of Natural Resources*. Merrill Publ. Co., Columbus, Ohio.

Houck, M. H., Cohon, J. L. & ReVelle, C. S. (1980) Linear decision rule in reservoir design and management 6. *Water Resources Research* **16**, 196–200.

House, P. W. & Mcleod, J. (1977) *Large Scale Models for Policy Evaluation*. John Wiley, New York.

Hwang, C. L. & Yoon, K. (1982) *Multi Attribute Decision Making: Methods and Applications*. Springer–Verlag, Berlin.

Iacobucci, E. E. (1985) *Application Display Management System*, IBM.

IAHR (International Association for Hydraulic Research) (1978) *Information Exchange on Computer Programs, edition 2*, July.

Inhaber, H. (1976) *Environmental Indices*. John Wiley, New York.

Jensen, S. (1983). Software and user satisfaction. In: Otway, H. J. & Peltu, M. (eds), *New Office Technology*, Chapter 11, pp. 190–204. INSIS Prog. Commission of the European Communities, London.

Keeney, R. L. & Raiffa, H. (1976) *Decisions with Multiple Objectives: Preferences and Value Tradeoffs*. John Wiley, New York.

Kindler, J. (1988) Systems Analysis and Water Resources Planning. In: *Proc. IFAC Symp. on Systems Analysis Applied to Management of Water Resources*, pp. 231–237.

Kohlaas, J. (ed.) (1982) *Compilation of water resources computer program abstracts*. US Committee on Irrigation, Drainage and Flood Control, Denver, Colorado.

Kraszewski, A. & Soncini-Sessa, R. (1984) *WODA — A Computer Package for the Identification and Simulation of a BOD-DO River Quality Model*, Clup, Milan, Italy.

Kraszewski, A. & Soncini-Sessa, R. (1986) WODA: A modelling support system for BOD–DO assessment in rivers. *Environmental Software* **1**, 90–97.

Krzysztofowicz, R. & Duckstein, L. (1979) Preference criterion for flood control under uncertainty. *Water Resources Research* **15**, 513–520.

Lee, N. (1983) Environmental Impact Assessment: a review. *Applied Geography* **3**, 5–28.

Loucks, D. P., Stedinger, J. R., & Haith, D. A. (1981) *Water Resources Systems Planning and Analysis*. Prentice–Hall, New York.

Loucks, D. P., Salewicz, K. A. & Codner, G. P. (1987) Dynamic simulation under uncertainty: An application to water resource systems planning. In: Lev, B. (ed.), *Strategic Planning in Energy and Natural Resources*. Elsevier Science Publ., Amsterdam, pp. 333–339.

Luenberger, D. G. (1979) *Introduction to Dynamic Systems*. John Wiley, New York.

Maguire, M. C. (1985) A review of human factors guidelines and techniques for the design of graphical human–computer interfaces, *Computers and Graphics* **9**, 221–235.

Major, D. C. (1977) *Multiobjective Water Resources Planning*, Water Resources Monograph No. 4, American Geophysical Union, Washington, DC.

Marglin, S. A. (1962) Objective of water resources development: A general statement. In: Maass, A. *et al.* (eds), *Design of Water-Resource Systems*. Harvard University Press, Cambridge, Massachusetts, pp. 17–87.

McBean, E. A., Hipel, K. W. & Unny, T. E. (eds) (1978) *Risk and Reliability in Water Resources*. University of Waterloo, Waterloo, Ontario.

Melli, P., Bolzern, P., Fronza, G. & Spirito, A. (1981) Real-time control of sulphur dioxide emissions from an industrial area, *Atmospheric Environment,* **15**, 653–666.

National Technical Information Service, US Department of Commerce (1977). *Computer programs in water resources*. PB-290 885, Office of Water Resources and Technology, Washington, DC.

Pagliari, M., Elisei, G., Frego, G. & Maini, M. (1984) Environmental telemetering: a pilot experiment in the management of thermoelectric plants (in Italian). *ENEL Rassegna Tecnica IV* **1**, pp. 40–47.

Peskin, H. M. & Seskin, E. P. (1975) *Cost–Benefit Analysis and Water Pollution Policy*. The Urban Institute, Washington DC.

Richardson, K. L. *et al.* (1980) *An inventory of selected mathematical models relating to the motor vehicle transportation system and associated literature*. UMI Research Press.

Rinaldi, S., Romano, P. & Soncini-Sessa, R. (1977) Parameter estimation of a Streeter–Phelps model. *J. Environ. Eng. Div. ASCE* **105-EE1**, 75–88.

Rinaldi, S., Soncini-Sessa, R., Stehfest, H. & Tamura, H. (1979) *Modeling and Control of River Quality*. McGraw–Hill, New York.

Ruzicka, R. (1988) Simul-R — A simulation language with special features for model-switching and analysis. In: *Proc. European Simulation Multiconf.*, Nice, 1–3 June 1988. SCS, Ghent, Belgium, pp. 28–32.

Schmidt, B. (1988). Systems analysis, model construction and simulation: the Simulation system SIMPLEX II. In: *Proc. European Simulation Multiconf.*, Nice, 1–3 June 1988. SCS, Ghent, Belgium, pp. 39–51.

Shriner, C. R. *et al.* (1978) *Inventory of data bases, graphics packages and models in the Department of Energy Laboratories.* US Department of Commerce, National Technical Information Service, Oak Ridge, Tennessee.

Streeter, H. W. & Phelps, E. B. (1925) A study on the pollution and the natural purification of the Ohio river. *Public Health Bulletin No. 146*, US Department of Health, Education and Welfare, Washington, DC.

Ushold, M., Harding, N., Muetzelfeldt, R. & Bundy, A. (1984) *Intelligent front end for ecological modelling.* Research Paper 223, Dept. of Artificial Intelligence, University of Edinburgh.

Yeh, W. (1985) Reservoir management and operation models. *Water Resources Research* **21**, 1797–1818.

Zangwill, W. J. (1967) Minimizing a function without calculating derivatives. *Computer J.* **10**, 293–315.

Zeigler, B. P. (1976) *Theory of Modelling and Simulation*, John Wiley, New York.

Zeigler, B. P. (1984) Stuctures for model-based simulation systems. In: Ören, T. I., Ziegler, B. P. & Elzas, M. S. (eds) *Simulation and Model-Based Methodologies: An Integrative View*, NATO ASI Series Vol. 10, Springer–Verlag, Berlin, pp. 185–216.

Zielinski, P., Guariso, G. & Rinaldi S. (1981) A heuristic approach to improve reservoir management: Application to Lake Como. In: *Proc. Int. Symp. on Real-Time Operation of Hydrosystems*, Waterloo, Ontario.

5

The role of artificial intelligence

This chapter gives an overview of principles and methods of artificial intelligence (AI) and of its specific importance in the design of EDSS. It starts with some basic techniques (an exhaustive and detailed discussion lies beyond the scope of this work), and continues with the integration of so-called knowledge bases and inference mechanisms with the other components of an EDSS. This chapter, which is an attempt to familiarize the reader with principles and basics of AI, is an introduction to the subject; therefore not much weight will be placed on formal presentations of logics and theorem proving.

The discussion about artificial intelligence (which may also be called machine intelligence) has been accompanying human science for centuries. People dreamed of building a machine that had some or all of the attributes of human capacities. With the invention of the computer, this dream seemed to be touchable. However, some questions, which accompany and form the basis for this discussion, have to be asked at this point: what are the human capabilities that make man intelligent, how are they characterized and, once determined, how can they be mapped into a machine?

John von Neumann (1966), a leading scientist who was also one of the cofounders of the computer science discipline, answered these questions by creating a scenario of machines which produce their own successors. These 'little' machines learn from the environment, and thus they increasingly enlarge their capabilities. This scenario seems to be a rather naturalistic and also idealistic picture of human society.

Other scientists, such as Alan Turing, rejected the question whether machines can think or not. Instead, he proposed the so-called Turing test for testing a machine for human features (Turing 1950). The experiment has to be carried out in the following configuration. There are three participants, a man (A), a woman (B) and a questioner (C). C sits alone in a room and is connected by Telex (or computer) to A and B. The task of C, who knows the others as X and Y, is to detect who is the man and who the woman. This is further complicated by the fact that A tries to deceive C whereas B tries to help him or her. What happens if the part of A is played by a computer? If the result of the test is the same, i.e. if C cannot distinguish between human and mechanical beings, it could follow that, machines are able to 'think'. Turing concluded that, at that time (the fifties), no computer had the necessary memory and calculating power to 'think', but he predicted that by the end of the century mechanical and cultural circumstances would have changed to such an extent that one could speak about thinking machines.

In the following years, attempts were made to construct general problem solvers, but these did not succeed owing to their too general approaches. In specific fields that are restricted by narrow boundaries successes have been achieved. One should remember that human flight was not the result of copying the motions and physical features of the bird, but was rather achieved by developing and using sources of energy and new techniques†. Today it is possible to build Expert Systems of surprising power in restricted areas, but it is not possible to represent general commonsense knowledge in a complete form, including contradictions, generality, reasoning about time, etc., as a human being uses it. These areas are subject to active research (McCarthy 1987).

In the following pages AI‡ will be discussed from an engineering point of view, following the generally accepted principles.Some important areas in which AI methods have proved to be useful and which constitute main research fields are the following:

- Natural Language Processing. Human beings communicate with each other using rather informal and hidden structures and also very context-sensitive terms or expressions. The machine, on the other hand, is an instrument with deterministic behaviour and fixed syntax. Bringing these 'systems' together is one of the main research areas.
- Expert Systems. Systems, based on logic and using rule-based deduction, have been constructed which provide humans with formalized expert knowledge in a wide range of fields (i.e. medicine, geophysics, etc.).
- Theorem Proving. This field was historically very important for the development of AI (Newell & Simon 1963): finding a theorem by using formal logical deduction and formalizing the skills of mathematicians is still an important topic.
- Robotics. This area is one of the central research fields, and not only because of its closeness to early dreams of mankind. Guiding and controlling robot actions is a complicated task, and is closely related to the so-called factory without human labour.
- Pattern Recognition (Vision). Making computers see and understand their environment involves a set of operations: sensing the environment, extracting basic features of the scene, and comparing these basic features with stored and comparable entities for the recognition process. More over, these entities could form part of a semantic model, giving some meaning to the recognized scene.
- Knowledge Representation. This involves mapping from real-world or human knowledge to machine representation. These techniques are also

† In the example only Input/Output relations are of importance, as in the case of a black box model. A different task would be to find a model which is able to formalize and to explain human cognitive processes.

‡ This term will be used throughout this chapter, even if the question about what is intelligent and whether machines can be intelligent or not remains open. This is due to the fact that in the computer science community this term is generally accepted.

fundamental in other areas and two examples will be described in the course of this chapter.

In the following sections, systems will be presented that are based on explicitly stored knowledge and the rules and frame-based approach to represent this knowledge will be described. Different possibilities for integrating such approaches with mathematical modelling will be discussed, as needed in a decision support framework. Finally, two examples of applications will be presented, one dealing with an object-based knowledge representation in the qualitative simulation of dynamic systems and the other using an Expert System for the analysis of river ecosystems.

5.1 RULE-BASED SYSTEMS

Before a description of the use of logic in AI is presented, some basic concepts need to be introduced (for a discussion see, for example, Nilsson 1980 and Hogger 1984). In many applications, such as Expert Systems, robotics, etc., a descriptive formalization of information is needed. First-order predicate calculus is a language which allows such a representation, and together with its compact and declarative representation capabilities, it contains a well-formulated set of logical features necessary for symbol manipulation.

Basic elements of the predicate calculus

The syntax of first-order predicate calculus and its symbols that will be used in this chapter are, very basically, as follows. Expressions which can be built with the symbols of this language and which are correct with respect to its syntax will be called *well-formed formulas (wff)*. The basic symbols of predicate calculus are predicates, variables, constants, functions and the so-called connectives. A *predicate* is used to describe a relationship within the universe of discourse. For example, to indicate that TICINO is a tributary of the river PO, the predicate TRIBUTARY could be used. An *atomic formula*

TRIBUTARY (TICINO, PO)

with the two terms TICINO and PO (also called constants) represents the fact that TICINO is a tributary of PO. *Constant* symbols are the simplest form of a term and describe distinct entities of the real world. *Variable* symbols, e.g. x, y, are also terms, but denote objects which are not defined at the beginning. So TRIBUTARY (x,y) would mean that x is a tributary of y, but x and y are not initially defined. *Functions* describe mappings in the domain of discourse. The real-world sentence 'the outflow of Lake Maggiore flows into the Ticino river, a tributary of the river Po' could be expressed by the following predicate calculus formula, where outflow is a function mapping LAKE MAGGIORE to a river:

TRIBUTARY (outflow (LAKE MAGGIORE), PO)

The following notation will be used here: capital letters denote predicates and constants, lower-case letters describe functions, and lower-case letters at the end of the alphabet indicate variables.

In each language, a meaning has to be correlated with the expressions that are constructed. In the present case, a wff can be given an interpretation by assigning a correlation between its elements and real-world entities and relations. This correspondence defines the semantics of the language. In predicate calculus, a formula has the value T if the corresponding statement about the domain is true. Otherwise, if it is false, it has the value F. Thus

TRIBUTARY (TICINO, PO)

has the value T and

TRIBUTARY (TICINO, NILE)

the value F.

Atomic formulas as described above may be combined by connectives: $\wedge$ (and), $\vee$ (or) and $\Rightarrow$ (implies). The connection of wff by $\wedge$ is called a conjunction; the single components' formulas are called conjuncts. The formula thus obtained has the value T if all conjunctives have the value T.

Formulas built by $\vee$ are named disjunctions, and their components are called disjuncts. A disjunction has the value T if at least one disjunct has the value T. The symbol $\Rightarrow$ is used to represent an 'if . . . then' conclusion; the formula constructed within it is called an implication, where the left-hand side is called the antecedent and the right-hand side the consequent. An implication is true if the consequent has the value T or if the antecedent has the value F. Finally, the symbol □. negates the value of a formula. A formula preceded by □. is called a negation.

Table 5.1 summarizes the calculation of the value of a formula, given the two basic wffs A and B.

Table 5.1 — Truth table

A	B	$A \wedge B$	$A \vee B$	$A \Rightarrow B$	□A
T	T	T	T	T	F
F	T	F	T	T	T
T	F	F	T	F	F
F	F	F	F	T	T

The connectives can be combined in different ways to form more complex wffs, and they may also replace each other. The basis for this process is formed by the following laws of de Morgan:

$\square (A \vee B) = \square A \wedge \square B$
$\square (A \wedge B) = \square A \vee \square B$

The implication A ⇒ B can be rewritten using □A ∨ B, as shown by the equivalence in Table 5.2.

Table 5.2 — Substitution of A⇒B by □A∨B

A	B	□A	A⇒B	□A∨B
T	T	F	T	T
F	T	T	T	T
T	F	F	F	F
F	F	T	T	T

The symbols presented up to now are sufficient to describe the so-called propositional calculus, where no variables are used in predicates. To use variables, constructs, which are called quantifiers, are needed. A formula consisting of a universal quantifier (∀x) preceding the formula P(x) has value T for an interpretation when the value of P(x) under this interpretation is T for all assignments of x to entities of the universe of discourse. A formula consisting of an existential quantifier (∃x) preceding P(x) has the value T if P(x) is T for at least one assignment of x.

The statement 'All children play with computers' might be represented as

(∀x) (CHILD (x) ⇒ PLAY-WITH (x,COMPUTER))

CHILD and PLAY-WITH are two predicates with obvious interpretations. Here x is the quantified variable and the scope of the quantification is the formula which follows the quantifier. x is also called a bound variable. Wffs where all variables are bound are called *sentences*. As another example, the real-world sentence 'There is a river which is a tributary of the river PO' might be written as follows

(∃x) TRIBUTARY (x, PO)

Unification

In working with quantified formulas, it is necessary to match the quantified variables, i.e. symbols that have not yet been defined must be given a specific value when proving a theorem or deducing new formulas. For example, if one wants to deduce from the two wffs P1(C) and (∀x) (P1(x) ⇒ P2(x)) the new wff P2(C), the constant C has to be substituted for the variable x. This process is called *unification*. Unification is of great importance in AI, as it allows a test to be made which checks if two expressions look alike.

The substitution can also be described as a set of $\{t_1/v_1, t_2/v_2, \ldots, t_n/v_n\}$. t_i/v_i means that the variable v_i will be replaced by the term t_i. Having the two wffs P(x,C) and P(B,y), a substitution {B/x, C/y} can yield P(B,C) and thus the two wffs can be unified.

Resolution

Before describing the *resolution* rule of inference, which allows new knowledge to be obtained from existing knowledge and which is the process of applying deduction rules to wffs for deriving new wffs, some remarks should be made about this process. If a wff has the value T under all interpretations, then the wff can also be called valid or a tautology. In a case where only constants are engaged, one only has to check with the help of the truth table if for all possible combinations the wff has the value T. If quantified variables occur, it is not always possible to decide whether a wff is valid. For this reason, predicate calculus is also called undecidable. But, if a wff is valid, then a procedure exists for verifying the validity of the wff. Furthermore, if the same interpretation gives each wff in a set of wffs the value T, then one can say that this interpretation satisfies the set of wffs. And a wff X logically follows from a set of wffs S if every interpretation satisfying S also satisfies X.

Resolution can be applied to derive new wffs from a set of wffs. In this process, a special form of wffs, the so-called *clauses*, is used. Clauses are wffs which contain only disjuncts. This is not a particular problem, since every wff can be transformed into a clausal form. In the resolution process, the two clauses to which the process is applied are called parent clauses, and the resultant clause is called the resolvent. Table 5.3 some cases of resolution.

Table 5.3 — Resolution (see Nilsson, 1980)

Parent clauses	Resolvent	Comment
Having the two clauses P and □P∨Q (P⇒Q)†	Q	*Modus ponens*
P∨Q and □P∨Q	Q	Q∨Q results in Q. (Merge)
□P and P	NIL	Empty clause or contradiction
□P∨Q (P⇒Q) and □Q∨R (Q⇒R)	□P∨R (R⇒R)	Chaining

† Alternative relations are shown in parenthesis.

Unification, which introduces the possibility of using quantified variables in the resolution process, will be demonstrated in the next example. This also exemplifies the use of resolution by refutation or contradiction. Suppose it is necessary to prove a wff from a set of wffs. A refutation system works as follows: first the goal wff is negated, then it is added to the other wffs and, subsequently, all are translated to clauses. In this set of translated wffs, resolution is used to derive a contradiction, i.e. the empty clause NIL. Following a contradiction from the original set of wffs and the negated goal, means that the original goal can be derived from our set of wffs and it is therefore proven.

(1) All people who write Expert Systems are scientists:

$$(\forall x)\ (E(x) \Rightarrow S(x))$$

(2) Dishwashers are not scientists: (?)

$$(\forall x)\ (D(x) \Rightarrow \Box\ S(x))$$

(3) There are dishwashers who make a lot of money:

$$(\exists x)\ (D(x) \wedge M(x))$$

Having these basic set of formulas, we want to deduce that

(4) There is someone who does not write Expert Systems and makes money:

$$(\exists x)\ (M(x) \wedge \Box\ E(x))$$

The negation of (4) yields $\Box\ (\exists x)\ (M(x) \wedge \Box\ E(x))$.

The single steps in normalizing wffs to produce clauses, which are introduced here rather informally, are the following, using the equivalences of the truth table presented above:

(a) Elimination of implications:

$$(1')\ (\forall x)\ (\Box\ E(x) \vee S(x))$$
$$(2')\ (\forall x)\ (\Box\ D(x) \vee \Box\ S(x))$$

(b) Reduce the scope of negation symbols, which means bringing negation into the formula so that it applies only to an atomic formula. The laws of de Morgan guide this process where the existential quantifier becomes a universal quantifier and vice versa:

$$(4')\ (\forall x)\ (\Box\ M(x) \vee E(x))$$

(c) Standardize variables, which means that in the scope of a quantifier one variable can be substituted for another without changing the value of that formula (i.e. the variables can be simply renamed). It is thus possible to ensure that each quantifier has its own unique variable.

(d) The existential quantifiers have to be eliminated. In case (3), the existential quantifier can be replaced by inserting a new constant C, which refers to an entity which is supposed to exist.

$$(3')\ (D(C) \wedge M(C))$$

(e) All universal quantifiers can be eliminated since all variables are bound. Thus there are only constants or variables which can be assumed to be universally quantified.

(f) Finally, the symbol $\wedge$ has to be eliminated. This can be done by replacing a wff $(P \wedge Q)$ by the set $\{P, Q\}$. This is possible because in both formal writings both wffs have to be proven to be true.

The original set of wffs now looks like a set of clauses

$$(1'')\ \Box\ E(x) \vee S(x)$$
$$(2'')\ \Box\ D(y) \vee \Box\ S(y)$$

(3a) D(C)
(3b) M(C)
(4″) □ M(z) ∨ E(z)

In order to prove the theorem, it is necessary to produce resolvents of the set of clauses and add these resolvents to the set. This process is continued until the empty clause is produced (see also Table 5.3).

(5) E(C) by (3b) and (4″) and substituting C for z
(6) S(C) by (1″) and (5) and substituting C for x
(7) □ D(C) by (2″) and (6) and substituting C for y
(8) NIL by (3a) and (7)

Now hypothetically, one could conclude from the set of logical formulas that there exist persons who make money and do not write Expert Systems (which is obviously true).

This method of deducing the contradiction is not the only valid one; there are several 'solutions'. And a chosen pair of parent clauses does not always result in a successful termination of the resolution process. In this case, the process has to go back to some previous choice of parent clauses (backtrack) and another choice must be made. In reality, there are several strategies for choosing a parent clause, and they are implemented in the inference process. These different strategies for choosing the parent clauses will not be discused here.

Production systems

Production systems are a special type of refutation system in which the wffs are categorized in two general classes: facts and rules. Facts are assertions about the real-world entities and are written in the form of predicates with the value T. Rules, on the other hand, are implications and state an 'if . . . then' sentence. In production systems, the formulas are not converted into the normal form as shown above. They remain in their original implicational description and are directly used in the resolution process. The base, which contains the set of facts and rules and which is similar to the set of original wffs in the former case, is used to derive a goal wff. In this base, knowledge about the real world is described in a declarative way, i.e. one describes what is true in the domain of discourse. The inference mechanism, as described before, works independently on this base. One does not describe how a problem is solved; rather, one declares what is true and the system can derive from this logically derivable knowledge. Production systems (see Fig. 5.1) are actually the most widely used technique for constructing Expert Systems.

This approach shows some advantages (Winograd 1975):

— Flexibility and economy: It allows the storage of knowledge in a compact way. It is easy to add new and to delete old knowledge. By stating a true fact or rule about the domain of discourse, the user does not anticipate how this knowledge will be used.
— Understandability: It is easy to understand the stated assertions and facts

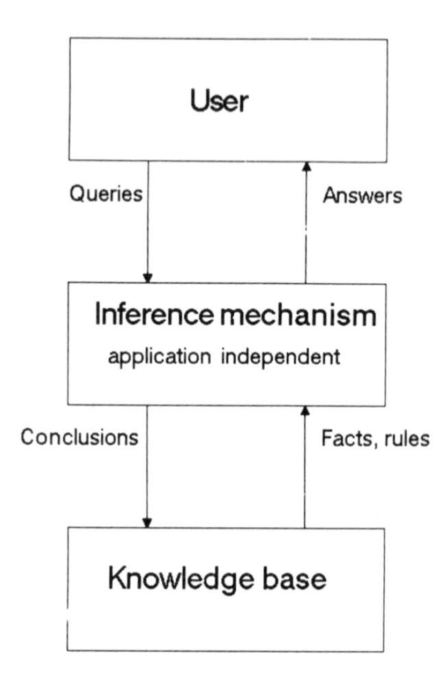

Fig. 5.1 — Architecture of a production system.

since both of them are kept in a short form and one only has to understand small and independent pieces of knowledge.

— Communicability: Since the system is also designed for non-experienced users, the set of facts and rules is easy to understand, and the compact description of the real world can serve as a communication and documentation tool.

Before describing in some detail a resolution process in a production system, which in substance remains the same as already presented, the notation will be slightly changed. The wffs will be called Horn clauses and have the form $P_0 \leftarrow P_1 \wedge P_2 \wedge \ldots \wedge P_n$, which means that to prove P_0 it is sufficient to prove P_1 and P_2 and . . . P_n. Only one predicate appears on the left-hand side and the right-hand side has the form of a conjunction. The symbol $\leftarrow$ can be interpreted as the $\Rightarrow$ (implication) symbol described above. All variables in the predicates are supposed to be universally quantified. Constructs, as above, are called rules, whereas implications with only the left-hand side of the $\leftarrow$ symbol are called facts. A clause in the form $P \leftarrow$ (right-hand side empty) simply states that P is true under all conditions. A negation is expressed by a clause with the left-hand side empty, i.e. $\leftarrow P$ states the negation of P (Genesereth and Ginsberg 1985). In the following example, well-accepted facts and rules about relationships are stated and some new fact is proved by using resolution.

(1) If x is the father of y, then x is also the parent of y:

$$P(x,y) \leftarrow F(x,y)$$

(2) If x is the mother of y, then x is also the parent of y:

$$P(x,y) \leftarrow M(x,y)$$

(3) If x is the parent of y and y is the parent of z, then x is the grandparent of z:

$$G(x,z) \leftarrow P(x,y),P(y,z)\dagger$$

(4) Girolamo is the father of Giorgio:

$$F(\text{Girolamo},\text{Giorgio}) \leftarrow$$

(5) Giorgio is the father of Gabriele:

$$F(\text{Giorgio},\text{Gabriele}) \leftarrow$$

(6) The proposition is now to prove that Girolamo is the grandparent of Gabriele, i.e. G(Girolamo,Gabriele), and for this one has to negate and to add to the set

$$\leftarrow G(\text{Girolamo},\text{Gabriele})$$

Unification proceeds as before except that, in the application of the *modus ponens* (see Table 5.3) used in the resolution process, the new form of describing rules has to be used. Having both rules $P \leftarrow Q$ and $R \leftarrow P,S$, it follows that $R \leftarrow Q,S$, which means that in the second implication the predicate P was replaced by its right-hand side from before, namely Q. This also describes the process of inference used in production systems: to prove the left-hand side of an implication, the single conjunctives at the right-hand side are replaced in a recursive process with their right-hand side expressions until the empty clause can be derived.

(7)	$\leftarrow P(\text{Girolamo},y),P(y,\text{Gabriele})$	from (6) and (3)
(8)	$\leftarrow F(\text{Girolamo},y),P(y,\text{Gabriele})$	(7) (1)
(9)	$\leftarrow P(\text{Giorgio},\text{Gabriele})$	(8) (4)
(10)	$\leftarrow F(\text{Giorgio},\text{Gabriele})$	(9) (1)
(11)	$\leftarrow$	(10) (5)

In step (7), (6) could be unified with (3) by replacing the variable x with the constant (real-world object) Girolamo. Furthermore, *modus ponens* produced (7) by substituting ← G(Girolamo,Gabriele) by (3). In step (9), the variable y was replaced by Giorgio and, finally, it was proved by contradiction that Girolamo is the grandparent of Gabriele, which is obviously true (for those who know Girolamo, Giorgio and Gabriele).

† Note that $\wedge$ is replaced in this notation by a comma.

5.2 FRAME-BASED SYSTEMS

Normally the domain-specific knowledge, which has to be represented as the base of an AI program, contains its own specific structure. Pieces of knowledge can be taken together to form a group or chunk of a specific description of a situation, an object or whatever one wants to deal with. This led to frames and frame-based systems (Minsky 1981, Bobrow and Winograd 1977). Such 'conglomerates' of information can also be described with the help of first-order predicate logic (Hayes 1979), but the AI community seems to have accepted frames and semantic networks as approaches that are better to understand and to use (Hendrix 1979, Quillian 1968).

Frame languages provide the knowledge base builder with an easy means of describing the types of domain objects that the system must model (Fikes and Kehler 1985). The description of an object might contain a prototype description of individual objects of that type; these prototypes can be used to create a default description of an object when its type becomes known in the model. Furthermore, heuristic knowledge can be added to that description (Winograd 1975), which seems to be hard to accomplish with a first-order logic approach. A frame provides a structural description of an object or a class of objects. Thus in Fig. 5.2, one frame represents a class of power plants whereas another describes a specific plant).†

The frame-based approach fits well with the object-oriented one presented in Chapter 3. Languages such as Smalltalk (Goldberg and Robson 1983) and LOOPS (Stefik *et al.* 1986) introduced the concept of objects also in software engineering. Although there are some differences between objects and frames, since objects can be seen as active entities which communicate by message passing and frames as a rather passive representation of information, in the following no distinction will be made between them.

Fig. 5.2 is a classification tree where thermal and hydro power plants are defined as kinds of power plants. PLANT XY is presented as a member of the class of nuclear power plants, as a distinct instance of one plant description. The different types of lines (solid and dashed) reflect this circumstance. This way of representing knowledge about real-world objects allows one to respect similarities between related objects (A is like B, with some exceptions) and also to introduce default assumptions. Some values are prewritten and might be overwritten by other frames at a lower level of the hierarchy.

Taxonomy and attributes

Frames provide a means of describing individuals and classes of individuals within the same framework. Frames are structurally organized in a taxonomy using two types of links between the single frames. The first type is a

† In the examples of this section, a notation similar to that of KEE (Fikes and Kehler 1985) is used; another language could be KRL (Bobrow and Winograd 1977). It should be noted that, since up to now no canonical form of frame language exists, concentration will be on the basic principles. KEE is a trademark of IntelliCorp.

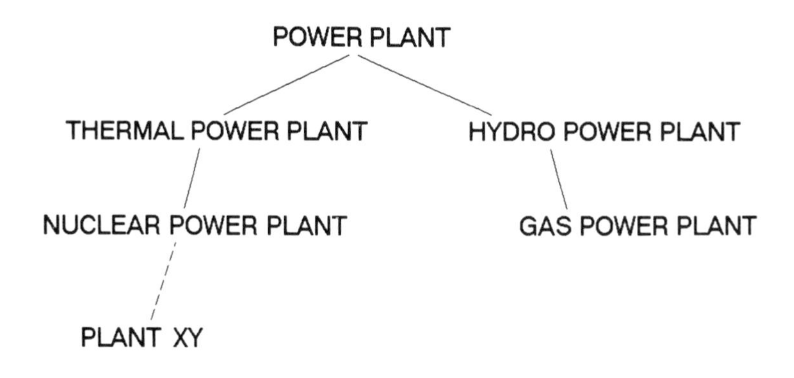

Unit:	THERM. POWER PLANT
Superclasses:	POWER PLANT
Subclasses:	NUCL. POWER PLANT
Member Slot:	PRODUCTION from POWER PLANT
Cardinality.Max:	1
Unit:	MW
Value:	Unknown
Member Slot:	TURBINES from POWER PLANT
Cardinality.Max:	10
Cardinality.Min:	1
Value:	Unknown
Member Slot:	OWNER from POWER PLANT
Value:	Unknown
Member Slot:	MEASUREMENT from TH. POWER PLANT
Cardinality.Max:	7
Cardinality. Min:	3
Value:	Unknown
Member Slot:	LOCATION from TH. POWER PLANT
ValueClass:	AREA
Cardinality.Max:	1
Cardinality.Min:	1
Value:	Unknown

Unit:	PLANT XY
Member of:	NUCL. POWER PLANT
Own Slot:	PRODUCTION from POWER PLANT
Cardinality.Max:	1
Unit:	MW
Value:	700
Own Slot:	TURBINES from POWER PLANT
Cardinality.Max:	8
Cardinality.Min:	1
Value:	6
Own Slot:	OWNER from POWER PLANT
Value:	ENEL
Own Slot:	MEASUREMENT fr. TH. POW. PL.
Cardinality.Max:	7
Cardinality.Min:	3
Value:	5
Own Slot:	LOCATION from TH. POWER PL.
ValueClass:	AREA
Cardinality.Max:	1
Cardinality.Min:	1
Value:	Town xy

Fig. 5.2 — Frame taxonomy.

subclass link, representing a specialization. The more generic description is kept on a higher level, and these types of links may also serve for introducing orthogonal concepts in the structural description in the domain of discourse. A thermal power plant can be seen as being orthogonal to hydro power plants. The second type of link is a member-of link, which reflects a class membership; a member of a class can also be interpreted as an instance of that type of class, e.g. PLANT XY in Fig. 5.2. (For a discussion of the different types of is-a links, see Brachman 1983.)

The description of the different attributes of a frame are contained in the so-called slots. In the specific language KEE, there are two types of slots: member and own slots. Own slots may be in any frame and describe attributes of that specific unit representing a class of objects or an individual. A member slot, on the other hand, stays only in the frame which presents a class of objects and serves to describe attributes of each member of that class. In Fig. 5.2, the member slot 'owner' reflects the fact that any specific power plant will be owned by some institution and thus has to be filled in from the single members of the class.

In the taxonomy tree formed by the frame structure in the domain of discourse, the single slots will be inherited in the subsequent frames. So the member slots of a frame describing a class of objects will become own slots of frames which are members of the class (see Fig. 5.2). On the other hand, in the case of subclass links, a subframe inherits the member slots of its superclass, which will be added to its member slots.† This feature of the frame approach also allows for efficient use of memory, since common attributes can be stored at the most central place, i.e. most generic frame, in the hierarchy.

As already mentioned, a single frame is described by its slots. Structural information on the single entities can thus be added easily in the appropriate types of slots. Moreover, a slot may have more than one value (for example the 'owner' slot of Fig. 5.2 might contain more than one value) and additional properties which are called facets. These specify the value of a slot and might also put some limits on that value. The facets Cardinality.Max and Cardinality.Min of Fig. 5.2 define the range of the possible values and can be used for proving the correctness of the results of a calculation. In frame languages, it is also possible for the value of a slot to be another frame. So the turbines could be described by frames which represent the different types of turbines. This feature allows for recursive nesting of memory structures, which are also close to a natural description of real-world objects.

Behaviour

The specific power of a frame-oriented approach arises from the fact that it allows, in addition to the structural description of objects, a procedural attachment to that object. In this way one can describe what happens when specific conditions are satisfied.

Two types of procedural attachments are provided by frame-based representation techniques: methods and active values (see Fig. 5.3). In most frame-based languages, these procedures are written in the host language (mainly LISP) and added in their original form to the frame description. Methods are stored in slots and act as message responders. Any time the

† Compare this concept with the similar one of the object-oriented approach with class and instance variables described in Chapter 3.

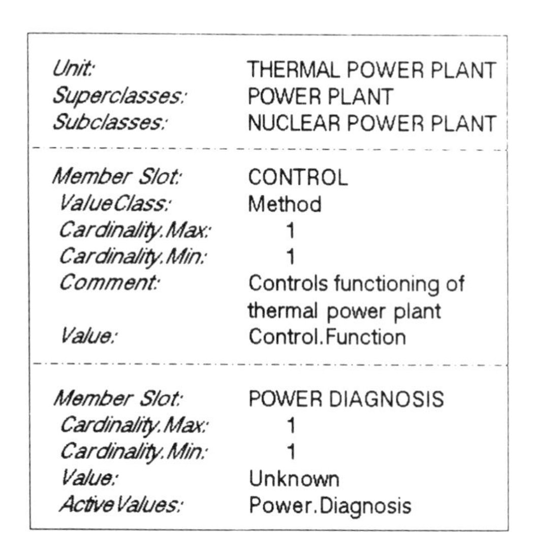

Unit:	THERMAL POWER PLANT
Superclasses:	POWER PLANT
Subclasses:	NUCLEAR POWER PLANT
Member Slot:	CONTROL
ValueClass:	Method
Cardinality.Max:	1
Cardinality.Min:	1
Comment:	Controls functioning of thermal power plant
Value:	Control.Function
Member Slot:	POWER DIAGNOSIS
Cardinality.Max:	1
Cardinality.Min:	1
Value:	Unknown
ActiveValues:	Power.Diagnosis

Fig. 5.3 — Methods and active values.

frame receives a message, the method identifies which slots are involved in the answer and provides the necessary calculations using the parameters supplied by the message. Active values, on the other hand, are functions that are correlated with a slot value and are called up whenever this value is read or written. So active values may control the correctness of the result of a calculation, or in the case of a power plant fault, could activate some messages to the operator or perform a primary diagnosis.

Reasoning

Frame languages also support reasoning about the information stored in the knowledge base. One can see that the knowledge structured in the form of frames is similar to the facts and rules in rule-based systems. As the latter have an incorporated inference mechanism, a similar mechanism can be provided in the frame approach. Disadvantages are that the formal basis of the deduction processes is not well developed, in contrast to first-order logic, and that up to now no standardized mechanism exists.

Since the representation of knowledge in the form of frames contains an inheritance structure, the reasoning mechanism can also attach to this structure. Similarities represented in the subclass links can be used to derive the fact that, in the case being studied, PLANT XY is a THERMAL POWER PLANT. The inference passes first a member-of link and then a subclass link. In addition, rules for deriving new facts can be described as frames. Using a frame, rules can be formulated in a more complex way. Such a frame could contain, besides the slots for the premise and assertion of the implication, slots for a description for the user, for an external representation, for a supplementary action which has to be processed when the rule is activated, or for describing the goals of the deduction for which this rule can be used. Such information can be utilized for choosing a specific sequence of rules in the resolution process. Moreover, the structure of frames provides the possibility of grouping rules to form a classification, which can be useful in building a modular rule base.

In the following example, a rule will be presented for classifying a Power Plant as one that has to be under careful control. A supplementary object, namely a SPECIAL CONTROLLED PLANT, which is defined as any thermal plant close to an urban area with a capacity of more than 300 MW, is introduced in Fig. 5.4.

Unit:	SPECIAL CONTROLLED PLANT
Superclasses:	POWER PLANT

and rule frame

Unit:	SPECIAL.CONTROL.RULE
Own Slot:	ACTION
Comment:	Action when rule is activated
Value:	Unknown
Own Slot:	ASSERTION from SPECIAL.CONTROL.RULE
Value:	(?X is in CLASS SPECIAL CONTROLLED PLANT)
Own Slot:	PREMISE from SPECIAL.CONTROL.RULE
Value:	(?X is in CLASS THERMAL POWER PLANT) (GREATERP (THE PRODUCTION of ?X) 300) (EQUAL (THE LOCATION of ?X) METROPOL)

Fig. 5.4 — Rule frame.

The predicates used above reflect the relations which can be described in such a language. The expression used in the value facet of the slot PREMISE has to be seen as a conjunction. As this frame language is embedded into LISP, the function (IN.CLASS) for class membership and any other LISP function, such as GREATERP, can be used.

Thus, consider the rule to say that PLANT XY is a SPECIAL CONTROLLED PLANT as (IN.CLASS PLANT XY SPECIAL CONTROLLED PLANT). The first conjunctive of the slot PREMISE matches the variable ?X with PLANT XY and returns the value T since, even if no explicit membership link exists between PLANT XY and THERMAL POWER PLANT, the rule interpreter sees that PLANT XY is a member of NUCLEAR POWER PLANT, which is a subclass of

THERMAL POWER PLANT. Thus, PLANT XY is also a member of the class THERMAL POWER PLANT. The second condition is fulfilled, since the PRODUCTION of PLANT XY is greater than 300. The last conjunctive involves looking at the slot LOCATION of frame PLANT XY. The value, stored in this slot, points to AREA frame TOWN XY, which is presumably of the type METROPOL. So the last condition also returns the value T, and it follows that PLANT XY is a member of class SPECIAL CONTROLLED PLANT. This causes a creation of the link member-of between PLANT XY and SPECIAL CONTROLLED PLANT. Furthermore, PLANT XY will inherit all member slots of the latter as own slots.

The two approaches, the frame-based and the rule-based, applied in both knowledge representation and inference, will be used in the following section for discussing possible combinations with mathematical modelling.

5.3 MATHEMATICAL MODELS AND AI

It is possible to integrate mathematical modelling and AI techniques as presented in the sections above, and such a combination can be seen as one of the fundamentals of the approach to building EDSS that is presented here. Before discussing integration, however, it is important to note the principal differences between mathematical models and AI approaches (see Table 5.4). The term mathematical models is used for 'classical' approaches

Table 5.4 — Some distinctions between mathematical models and AI methods

Mathematical models	AI
number crunching	symbol manipulation
procedural description	declarative description
how a problem is solved	what is true in the problem domain
trajectories/optimum calculation	goal deduction
explicit time dimension	static description
	modular enrichment
	independent resolution mechanism

in areas such as optimization, simulation, etc. (as presented in Chapter 4), and AI for knowledge representation techniques and inference mechanisms. Although the description of a problem and its solution, for example in the form of a dynamic simulation program, can be seen as knowledge representation, the use of this term is limited to the techniques described in sections 5.1 and 5.2.

Historically and methodologically the first relationships between the two areas can be drawn between simulation, one of the classical areas of mathematical modelling, and artificial intelligence methods (Birtwistle 1985, Kulikowski 1987). The latter can support the first in several ways:

— Automatic program generation, which means that the coding in the form of a computer program of a formally well-defined system to be simulated is fulfilled by the machine (Mathewson 1975).
— Support in the construction and selection process of a model where structural and semantical properties of the modelled systems are managed and controlled automatically by the machine (Zeigler and De Wael 1986, Guariso *et al.* 1987, 1988, Ören 1984).
— Help in the interpretation of results of simulation runs where statistical methods and graphical representation techniques are automatically or semi-automatically applied. These may also be viewed as intelligent front ends (Bundy 1985).
— The simplification process in existing models can be supported. For example, in the case of dynamic systems, this can lead to models with fewer state variables and a simpler structure (McRoberts *et al.* 1985).
— In complex systems, for example simulation with multiple decision making agents, heuristic and symbolic subprocesses (for example hypothesis formulation) could be formulated with first-order logic, whereas subprocesses that are continuous and deterministic (such as physical processes) could be presented as dynamic models (Kerckhoffs and Vansteenkiste 1986).

On the other hand, simulation produced and still produces benefits for research in AI, especially in the field of knowledge representation. The object-oriented technique in software engineering, a subject related to the frame-based approach in AI, has the simulation programming language SIMULA as one of its ancestors (Dahl and Nygaard 1966). In SIMULA, which was developed in 1966, the concept of objects and classes with the inheritance mechanism is already present.

The problem of time

Simulation provides a straightforward approach to representing time, up to now a crucial problem in AI. AI programs have difficulties in dealing with past and future events, i.e. in saying what is actually true and what can be true in the future, and what action at what time will change the state of a system. Different actions have to be performed with respect to different states of a system, and knowledge bases have to be updated with respect to time and its change.

From the fact 'x is writing an article' one can conclude that 'x is a diligent guy'. If x is going to have lunch, to take a holiday or to hold lessons, the first fact has to be eliminated, but the conclusion remains the same. A proper inference would be from 'If x has written an article recently, he is a diligent guy' and 'x is writing an article' to 'For the next time, x will have written an article recently' and hence to 'For the next time, x is a diligent guy'. (What will happen if x does nothing more?) In McDermott (1982), a readable and already classical description of the problem is given and a temporal logic for reasoning about time is proposed. He introduces the concept of subsequent states of the domain of discourse and their ordering with respect to time. Furthermore, he defines chronicles which are totally ordered sets of states and represent possible histories of the universe. His language is based on first-order logic.

Qualitative modelling

Another approach to reasoning about time is used in the so-called qualitative modelling concept (Forbus 1984), which is based on a frame technique and shows a closer relationship to simulation. In Rajagopalan (1986), an overview of different techniques used in qualitative modelling and simulation is given. Whereas in quantitative modelling one has to know the exact values of all parameters, the description of the system by means of equations has to be complete, and sometimes quite complex solution procedures have to be used (which are not very close to everyday knowledge), qualitative modelling tries to represent commonsense knowledge about the physical world with its possible changes, histories and processes. This includes simple descriptions of cause – effect relationships between variables, which are described in qualitative terms. Magnitudes of quantities are defined in ordered relations between these quantities and their direction of change (derivatives). An algebra is developed for quantities described in that way. Time is introduced as events, which change states, and as histories. Qualitative modelling allows the consideration of the following tasks:

— Determining activity: What is happening in a situation at a particular time?
— Prediction: What will happen in future in some specific situation?
— Postdiction: How might a particular state have come about? There is broken glass on the floor. Going back in time, one could deduce some events in the history and maybe see a bottle falling.
— Sceptical analysis: Determining if the description of a situation is consistent.
— Measurement interpretation: Having a description of individuals in a specific situation, one could infer what else is happening.
— Experimental planning: Knowing what can be observed and manipulated, planning action will yield more information about the situation.
— Causal reasoning: Supposing causality in the system to be simulated, causal relations in changing attributes and behaviour can be deduced.

The process of heating water can be used as an example for demonstrating the qualitative modelling and simulation approach. First, the temperature increases, and after some time the water begins to evaporate, representing a change of state from liquid to gaseous (a further possibility might be that, once water begins to evaporate, the amount of water decreases and it may vanish). Simulation of this process continues until a recognized pattern of behaviour occurs (stable state, oscillation, etc.). The biggest problem in this approach is constituted by the ambiguities which might arise during the calculation with the qualitatively described quantities. What happens, for example, if two variables with nearly the same magnitude are subtracted from each other: will the result be zero? In such cases, rules about the specific problem domain may help, or simulation has to continue examining all alternatives.

The basic simulation process, proceeding over time, passes the single states of the systems, examining the possible changes of the qualitative variables. These changes identify the set of possible subsequent states, and more than one future alternative may arise. It may also be possible to backtrack and to follow another alternative in an ambiguous situation.

The qualitative modelling approach supports the model user by giving him or her the possibility of getting explicit causal information from the model while following its dynamics. This cannot be accomplished in a classical quantitative simulation program, which produces only state variable trajectories, and possible correlations may emerge only through the specific skills of the person interpreting the results.

5.4 INTEGRATING MATHEMATICAL MODELS AND AI METHODS

At the implementation level, there are three tendencies in the attempt to integrate or to combine AI methods with mathematical modelling. As already mentioned, experiments are mainly correlated with simulation, but can also be found in other fields. These three directions are:

— frame- or object-based environments,
— model calls as operational predicates in first-order logic, and
— AI languages (for example Prolog) for formulating mathematical (simulation) models.

Object-oriented environments

The object-oriented approach supports the structural representation of knowledge in the form of objects as well as the declarative and procedural description of a solution mechanism. This attractive method has already produced a certain number of simulation environments (Bezivin 1987, Klahr and Faught 1980, Kerckhoffs *et al.* 1986, Luker and Adelsberger 1986). The simulated real-world objects might be described as programmed objects guaranteeing a rather complete mapping of the real entities (see also sections 3.5 and 5.2). Languages such as Smalltalk and LOOPS may be used for knowledge representation, database management and model formulation. This supports an integral implementation of a whole system without using different languages for components. Such systems allow fast prototyping and the creation of models and their simulation. Moreover, interactivity allows the fast updating of model parameters and also model structure, thus shortening the time for model construction and testing. Additional stored structural information assists in the correct formulation of a model. The prototype EDSS described in Chapter 7 relies heavily on this approach.

Model calls within first-order logic

The central idea of this approach is that specific and well-marked predicates of first-order logic may represent calls of external models, while their arguments are the input and output parameters of such a model. This allows mathematical models to be integrated into first-order logic, provided some administrator may guarantee the correctness and executability of the external models. Thus, the use of mathematical models can be controlled by knowledge stated in the form of rules. A user is thus supported in the selection and proper use of a model.

For the purpose of the present discussion, it is assumed that external model calls return the boolean value T if the calls were successful and the output of the models were calculated correctly. In addition, the concept of preconditions is introduced. Predicates representing external models can only be evaluated (i.e. called) if all input parameters are instantiated, i.e. all variables are replaced in the matching process. The inference mechanism remains the same as presented in section 5.2. Similar approaches are proposed in Bonczek *et al.* 1981 and Dolk and Konsynski 1984.

In the following example, a modified version of one presented in Bonczek *et al.* 1981, the predicate PREDICT($\underline{y}$,$\underline{x}$,r) represents a call of a model where the arguments y and x have to be instantiated prior to the call. The predicate returns the value T if the value of the argument r is calculated correctly. From now on r can also be used as an instantiated variable. The following example demonstrates the use of such operational predicates inside first-order logic and the resolution process to deduce new information.

Suppose that there are four models: a regression model calculating a regression coefficient; a prediction model for forecasting toxic contents having a production index and a regression coefficient; a model which computes the concentration of pollutants based on the amount of emission and the dimension of the geographical area considered; and a model calculating the emission as a function of the technological efficiency of a plant and the toxic contents of its input. The respective predicates are REGRESS, PREDICT, CON_CALC and EM_CALC. Furthermore, there are predicates (representing time series) for the yearly amount of production in a specific factory (PROD), yearly production indices (PRO_IND), emissions (EM), concentration (CON), toxic contents (TOX), technological efficiency (TECH), technological indices (TEC_IND), and predicates indicating the year (YEAR) and the surface area (GEO). These predicates may be stored in a database where they can be retrieved. All variables are universally quantified.

Clause (1) below states that if x are indicators of the production in a factory in year yr_1, m are the indicators of another year yr, and y is the real production, both y and x are arguments for a regression model which produces φ as an output, and a prediction model which calculates the toxic contents p in production, then the toxic amount in the products of the year yr is p. Or, in other words, having historical data about production and production indices, the toxic contents of a factory's products can also be predicted for a year where complete data are not available.

Clause (2) represents the fact that, if there are emissions em in year yr on a certain area pa and a model CON_CALC which calculates from em and the area pa the concentration con, then there is a concentration con in the area at year yr. Clause (3) is a formulation similar to (1), but it deduces from efficiency in technology (TECH) in a specific year and from technological indices (TEC_IND) the efficiency in technology in a new year. This efficiency serves as a parameter for the filtering of toxic substances. Clause (4) states that if there is technological efficiency t and toxic content p and a model EM_CALC that calculates from t and p the emission em, then there is an emission em in year yr. Clauses (2) and (4) state that if there are data about technological efficiency and toxic contents, one can first deduce the amount of emission and, with the information about the surface area, the concentration in each year.

Thus, the four clauses with their incorporated models serve to predict emissions and concentrations, presuming that the respective technological and production indices are available. (As an exercise, the reader may convert these clauses to their original implicational form.)

(1) $\square REGRESS(\underline{y},\underline{x},\varphi) \vee \square PREDICT(\underline{\varphi},\underline{m},p) \vee \square PROD(y,yr_1) \vee$
$\square PRO\_IND(x,yr_1) \vee \square PRO\_IND(m,yr) \vee TOX(p,yr)$

(2) $\square EM(em,yr) \vee \square GEO(pa) \vee \square CON\_CALC(\underline{em},\underline{pa},con) \vee$
$CON(con,yr)$

(3) $\square REGRESS(\underline{v},\underline{u},\tau) \vee \square PREDICT(\underline{\tau},\underline{n},t) \vee \square TECH(v,yr_2) \vee$
$\square TEC\_IND(u,yr_2) \vee \square TEC\_IND(n,yr) \vee TECH(t,yr)$

(4) $\square TECH(t,yr) \vee \square TOX(p,yr) \vee \square EM\_CALC(\underline{t},\underline{p},em) \vee. EM(em,yr)$

(5) $\square CON(con,yr) \vee YEAR(yr)$

(6) $\square CON\_CALC(\underline{em},\underline{pa},con) \vee GEO(pa)$

The last two clauses serve in the following deduction process. The goal as stated is to find the concentration c in year 1988 on a surface of 50 units $(\exists C(CON(c,yr) \wedge YEAR(1988) \wedge GEO(50)))$. For this year, only production and technological indices are available. One would then negate this formulation, add it to the set and try to find a contradiction (i.e. the clause NIL) in the rule base.

(7) $\square CON(c,yr) \vee \square YEAR(1988) \vee \square GEO(50)$

with clauses (7) and (5), 1988 goes for yr and c for con

(8) $\square CON(c,1988) \vee \square GEO(50)$

with (8) and (2), c for con, 1988 for yr, 50 for pa

(9) $\square EM(em,1988) \vee \square GEO(50) \vee \square CON\_CALC(\underline{em},\underline{50},c)$

with (9) and (6), c for con, 50 for pa

(10) $\square EM(em,1988) \vee \square\ CON\_CALC(\underline{em},\underline{50},c)$

with (10) and (4), 1988 for yr

(11) □CON_CALC(em,50,c) ∨ □TECH(t,1988) ∨ □TOX(p,1988) ∨ □EM_CALC(t,p,em)

with (11) and (1), 1988 for yr

(12) □CON_CALC(em,50,c) ∨ □TECH(t,1988)∨ □EM_CALC(t,p,em) ∨ □REGRESS(y,x,φ) ∨ □PREDICT(φ,m,p) ∨ □PROD(y,yr_1) ∨ □PRO_IND(x,yr_1) ∨ □PRO_IND(m,1988)

with (12) and (3), 1988 for yr

(13) □CON_CALC(em,50,c) ∨ □EM_CALC(t,p,em) ∨ □REGRESS(y,x,φ) ∨ □PREDICT(φ,m,p) ∨ □REGRESS(v,u,τ) ∨ □PREDICT(τ,n,t)

(14) ∨ □PROD(y,yr_1) ∨ □PRO_IND(x,yr_1) ∨ □PRO_IND(m,1988) ∨ □TECH(v,yr_2) ∨ □TEC_IND(u,yr_2) ∨ □TEC_IND(n,1988)

The model calls are obtained in (13) and the other predicates in (14) which in the present case stand for queries for data stored in a database. If, in asking for example TEC_IND(n,1988), the index N_0 is found, then the variable n will be replaced by the constant N_0 and the predicate is evaluated to T. Applying this procedure to all predicates in (14), the data are obtained for the input arguments of (13). Thus, this clause can be rewritten as:

(13) □CON_CALC(em,50,c) ∨ □EM_CALC(t,p,em) ∨ □REGRESS(Y_0,X_0,φ) ∨ □PREDICT(φ,M_0,p) ∨ □REGRESS(V_0,U_0,τ) ∨ □PREDICT(τ,N_0,t)

now calling twice REGRESS, substitute Z_0 for φ, S_0 for τ

(15) □CON_CALC(em,50,c) ∨ □EM_CALC(t,p,em) ∨ □PREDICT(Z_0,M_0,p) ∨ □PREDICT(S_0,N_0,t)

executing twice PREDICT and T_0 for t, P_0 for p:

(16) □CON_CALC(em,50,c) ∨ □EM_CALC(T_0,P_0,em)

calling EM_CALC we get EM_0 for em:

(17) □CON_CALC(EM_0,50,c)

and finally, calling the last model, one obtains the value of c and

(18) NIL

In the deduction from (13) until (18), the calls of the respective models result in evaluated predicates with the value T so that the resolution can go on. As a final result, one obtains the variable c instantiated by a value which represents the calculated value of the concentration in 1988.

AI languages for formulating models

In contrast to the previous approach, AI languages, typically Prolog (Clocksin and Mellish 1981), may be used to formalize models, mostly simulation models. Prolog is based on first-order logic and works like a production system (see section 5.1). The sequence of inference is defined by the textual order of the rules. Extensions of Prolog such as Concurrent Prolog (Shapiro and Takeuchl 1983), TS-Prolog (Gergely and Futo' 1983) and T-CP Prolog (Cleary *et al.* 1985) reflect the need to combine AI with simulation. In contrast to sequential Prolog, where the sequence of the inference is determined by a fixed strategy, in Concurrent Prolog and T-CP Prolog the next rule or fact to be taken as parent clause is chosen randomly. Moreover, multiple subgoals may be under consideration at a certain moment, allowing parallel deduction, for which supplementary synchronization tools are provided. A typical clause in T-CP Prolog looks like

$$A \leftarrow G_1, G_2, \ldots, G_n < \text{:delay:} > B_1, B_2, \ldots, B_m$$

The G_i are called guards and have to be evaluated in the inference steps before the goal A may be replaced by the subgoals Bj. The choice of goal A out of a set of possible goals is random and may occur more than once until all G_i are evaluated to be true. Inference with this rule can only proceed after the time 'delay' has passed. In other words, A can be replaced by the B_j if all G_i are true and time delay has passed. By this, concurrent goals can be synchronized, and the non-deterministic choice of the next rule together with the waiting process guarantees that the rule with the minimum delay will be chosen. A global simulation time is maintained by the language.

T-CP Prolog is well suited for discrete event simulation, and programming in such a language allows good integration of simulation with knowledge representation and reasoning techniques in rule form.

5.5 THE ROLE OF AI IN EDSS

In some sense DSS, or as in the present case their specific application to environmental problems, can be seen as one of the attempts to integrate developments in AI with those in other areas. This is especially true because DSS are the product of an interdisciplinary attempt at integration, not only of different applicative engineering disciplines but also of several fields of computer science. Within these systems an inexperienced user should have the possibility of using and creating a wide range of different mathematical models or other tools, such as a database, that could support the problem solving. Furthermore, these components have to be maintained and augmented by new features, if necessary. To guarantee this flexibility, user friendliness, modularity and adaptability of the system, the support of AI techniques seems to be important.

A possible means of integration could be the insertion of AI techniques into the single components of a simple EDSS, as suggested in the tool-based approach of section 1.5. They contribute to the functioning of the single modules of this kind of DSS, i.e. database, modelbase and dialogue part, and the conceptual architecture remains the same. Another possibility is to add supplementary components, such as the knowledge base and partly also the system management unit (see also Elam and Henderson 1983), and to augment the other components to be able to have access to the information in this 'intelligent' store. Before describing the possible contributions of inference and knowledge representation to EDSS, some distinctions are drawn between features of EDSS in general and Expert Systems, as a well-known representative of AI approaches (see also section 5.1 about production systems). On a conceptual level, one can see the following difference: EDSS tend to prepare a user for a decision, whereas an Expert System tries to replace the human expertise or expert, and thus takes the decision itself (see Table 5.5).

Table 5.5 — Distinctions between DSS and Expert Systems as a representative of AI tools (derived from Turban and Watkins 1986)

Attributes	DSS	Expert systems
objectives	assist in decision making	copy and replace human capabilities
major orientation	decision making	transfer of expertise and rendering of advice
who makes decision	user and/or system	system
problem area	broad, complex	narrow
world view	open	closed
query direction	human queries	system queries human
data manipulation	numerical/symbolic	symbolic
mathematical models	incorporated	—

The following list is an overview of the different tasks that could be managed by the methods presented in the previous sections. This list contains possible contributions within both approaches that have been mentioned: integration of AI methods into the single modules of an EDSS and the addition of a separate knowledge base. These contributions have to be seen as a description of possible features, without details on how they can be implemented.

— A knowledge base could contain the structural knowledge about the stored models (i.e. their types, internal structure, information how to build them, etc.). This information is needed for the construction of a mathematical model when no proper candidate can be found in the modelbase to satisfy the user's needs. A prototypical example of this usage will be presented in Chapter 7.

— Using a frame-based approach for the storage of the structural information, the instances of the classes which describe the types of models can be used to represent the distinct models, i.e. the models that really exist in the system in an executable form.

— Frames, which describe models, or rules may contain information that allows a user to choose a model, to detect the specific aim of a model and to use it.

— The model-specific information in a knowledge base might be of importance in the case of an overall inference mechanism that includes model calls (see also section 5.3). Here, the choice of rules used in the inference could depend on the deducible goal.
— A knowledge base may store information on the correct ordering of models when they are executed in sequence. Additionally, it should guarantee that the intermediate data are stored in the database to be used by succeeding models. This also includes the management of different levels of memory with respect to the actuality of data. In this sense the intelligent module should fulfil the tasks of a database and a modelbase management system.
— Whereas the tasks described up to now deal with general knowledge, there should also be the possibility of incorporating domain-specific information, i.e. knowledge about the problem area. This might be included in the form of data in the database but also in the form of rules or frames. This information is needed, together with the structural and instantiated knowledge about models, for the construction of new models and the usage of existing models.
— As the success of an EDSS depends much on the interface and the interactivity, modelling a profile of a user or of classes of users could guide the work with the systems and its adaptation or the training of new users.
— The overall structure of the system with its different components, stored as some meta-information, should also be the object of the knowledge processing, providing reasoning about itself and control of the system.
— The modelbase could contain qualitative models. These could be used for hypothesis formulation or for examining rapidly the qualitative behaviour of a real system, where later on the interesting parts can be simulated with quantitative models to obtain exact numerical results. This strongly supports human intuition in the formulation of problem solutions.

The following two sections exemplify possible means of integrating knowledge representation and inference mechanism with mathematical modelling. Both examples arise from the field of simulation, thus reflecting the fact that the proposed integration has already shown very positive results in this area.

5.6 AN EXPERT SYSTEM FOR ANALYSING RIVER ECOSYSTEMS

CARE (De Filippis 1988), the acronym for Computerized Analysis of Riverine Ecosystems, is an Expert System which helps the user to determine whether the features of a given river course (whose main purpose is, for instance, to supply agriculture or hydropower production) can sustain a certain fish population.

The system is a simplified version of the IFIM method (Bovee 1982), and it is limited to the analysis of a single fish population. The population dynamics is represented by a four-state-variable linear model, where each variable represents the number of fish at each stage of life. More precisely, the equations describing the development of the population are as follows:

$$S_{t+1} = k_e A_t/2$$

$$F_{t+1} = k_s S_{t+1}$$

$$J_{t+1} = k_f F_t$$

$$A_{t+1} = k_a A_t + k_j J_t$$

where:

the unit of time is one year,
S represents the number of spawning fishes,
F the number of fry,
J the number of juveniles, and
A the number of adults.

The coefficient k_e represents the average fecundity of a spawning female, while k_s, k_f, k_a, and k_j represent the survival rates between one stage of life and the next. All these coefficients depend upon the fish species being considered.

The IFIM method is based on the assumption that there is a minimum wetted area in the river required to provide a sufficient habitat for each fish species and for each life stage. The total available wetted area can be computed on the basis of geometrical and hydraulic characteristics of the stretch of river considered. Furthermore, other environmental factors may modify the habitat suitability. In particular, it is assumed that each of these factors can be characterized by an optimal value and by threshold values outside which the development of the population is prevented (see, for instance, the temperature preference curve in Fig. 5.5). Between the optimal and the extreme values, there is a decrease in the habitat suitability, and this decrease can be represented as a contraction of the area available for the fishes and thus a decrement of the sustainable population. Among the environmental factors affecting habitat suitability, the system considers oxygen and pollutant concentrations, type of river bed, water temperature and river stage.

All this information is stored in the system in the form of rules which form the knowledge base of the system. Starting from a set of known facts (for instance the fish species and the population size) an inference mechanism, i.e. a procedure that tries to use the rules to assess new facts as shown in section 5.3, allows one to determine whether the target population may survive in a given water course. If it is found that the desired population cannot be sustained, the bottlenecks are determined.

All the models for assessing the suitability of the habitat and its value are implemented as separate external programs (written in a language such as

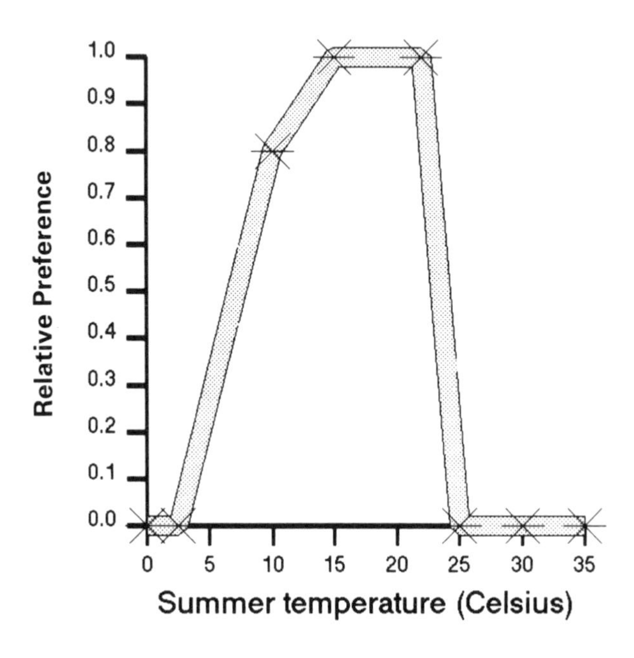

Fig. 5.5 — A sample temperature preference curve.

Pascal), which are invoked as consequents of certain rules. For instance, fish may survive only if the oxygen concentration is sufficiently high; thus, if pollutant discharges are present along the river, a model must compute in which portion of the stretch of river under consideration the dissolved oxygen falls below the critical threshold. This is accomplished by an external program which simulates the Streeter–Phelps model (see section 4.6.1 for details). This computes the concentrations of pollutant (Biochemical Oxygen Demand BOD) and dissolved oxygen (DO) in the stretch, to determine which portion of the stretch presents suitable habitat conditions.

To assess the suitability of the environment, the system thus runs, in a coordinate sequence, a number of mathematical models describing, besides the dynamics of pollution degradation along the river, the geometry of the river and the preferences of the fishes at the different life stages (see Fig. 5.6). Each of these models is represented by a separate external program, which is invoked as a consequent of some rule in the knowledge base. The system then interprets and combines the results of all these models to answer such questions as

— Given a specific habitat, will the desired population survive?
— What are the reasons for not succeeding?
— What happens if there is a change in:

the water flow-rate,
the geometry of the river,

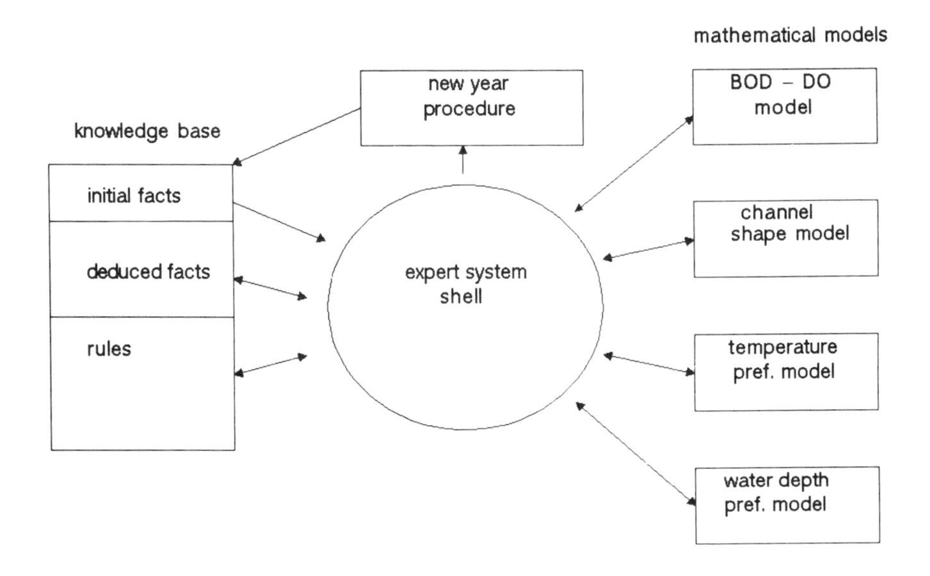

Fig. 5.6 — Architecture of the CARE Expert System.

the amount of pollutant discharged, or
the mortality or fertility rate?

The most interesting feature of the system is that a simulation of several years is made possible by modifying the fact base from which the inference engine starts the searching procedure. In this way, time-dependent behaviour can be taken into account. This is unusual in Expert Systems, where the fact base normally contains static knowledge.

If the user wants to continue the analysis for a new period, another program is called by the system (see again Fig. 5.6) and writes the relevant data onto an external file, from which they may be read in order to continue the simulation. The following two rules present the calculation of the numbers of fish for the following period and the call of the program to update the base of initial facts.

IF the habitat is sufficient THEN
  number of spawnings = k_e * number of adults/2 AND
  number of fry = k_s * number of spawnings AND
  number of juveniles = k_f * number of fry AND
  number of adults = k_a * number of adults + k_j * number of juveniles

IF the user wants to simulate another period THEN
  CALL 'new_year_procedure' (length of river stretch;
    river depth; width of river; type of river bottom;
    desired number of adults, juveniles and fry; actual
    period; actual number of adults, juveniles and fry;
    degradation and reoxygenation coefficients; >next period<)

In the first rule, the premise 'the habitat is sufficient' has to be deduced before computing the new fish population. In the second, the arguments of new_year_procedure represent facts which do not change in time, and thus must be rewritten into the fact base, and are necessary to analyse the new fish population. The last argument, marked with > <, is the result of the call. All values are written onto a file, and the inference mechanism restarts.

CARE represents an interesting attempt to combine mathematical models and Expert Systems into a unique package. In this way, it enjoys the precision and speed of quantitative evaluations performed by separate programs, which represent single aspects of the overall problem, together with the overall coordination, the explanation and dialogue facilities of an Expert System. The user can thus pose his or her questions in a plain English style and with several different common formats (a dictionary of synonyms provides the necessary translation). He or she can ask the system how each conclusion has been reached, and the responses of the system are in a clear textual form. The application of simulation models and other software remains completely hidden to the user, who is only asked for the necessary data. Even during this phase, the Expert System assists the user by performing a check on the entered inputs, supplying upon request additional explanations of the type of data, measurement units, sampling procedures or explaining why a certain value is needed. This method of integrating AI tools and simulation software seems to be one of the most promising approaches to developing environmental software (Guariso 1988).

5.7 AN EXAMPLE OF QUALITATIVE MODELLING AND SIMULATION

A qualitative model builder and simulator is presented as the second example. It is a description of a prototype, developed for testing several strategies connected with problems in qualitative simulation, i.e. for solving ambiguities. It fits well into the present overall framework, since such a qualitative simulator may support a decision maker with a first approximation of his or her problem. Moreover, it allows for very fast model construction, needing only a few minutes for a problem of medium complexity (the aspects and details of its interface design will be described in Chapter 6).

The system is based on an object-oriented point of view and is written in LOOPS. The user states the problem using qualitative terms (i.e. big, small, etc.) for its variables and their relations. Variables are created, and their attributes (i.e. magnitude, sign, etc.) are described using the graphic interface facilities of the system, such as windows or the mouse. Relationships are assigned by connecting graphically the two respective variables, represented as boxes, on the screen with the help of a mouse. The only two things a user has to be acquainted with are the concept of a dynamic system with input, state and output variables and some interface basics as how to use a mouse or the keyboard.

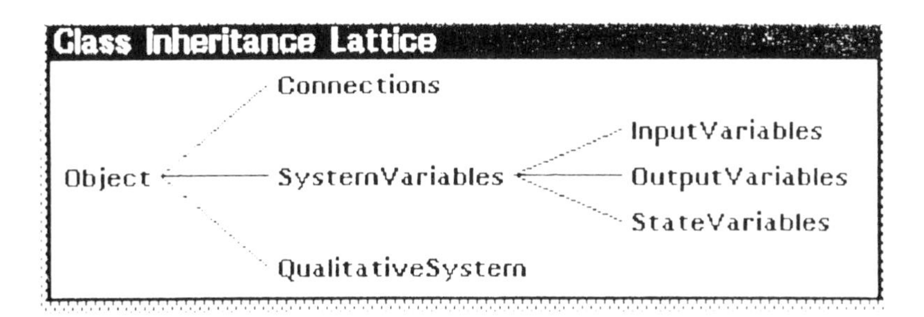

Fig. 5.7 — Hierarchy of classes in the simulator.

Fig. 5.7 represents the hierarchical structure of classes in the prototype. For implementational reasons, the class object is inserted as a root into the structure. The real objects, created by the user, in the qualitative system are instances of the leaves of the inheritance lattice. The three types of variables in a dynamic system have a common superclass because of attributes they possess in common (see Fig. 58).

The simulated system is described by an instance (realization) of the class QualitativeSystem, which possesses as instance variables the lists of the respective variables and connections. It should be noted that there is a difference between the static class hierarchy, where connections and variables are on the same level as QualitativeSystem, and an instance of the class QualitativeSystem, which possesses instances of the classes InputVariables, StateVariables and OutputVariables for describing its own structure.

The class SystemVariables has lists for its predecessors and successors as well as a description of its magnitude and sign. These are properties which are common to all variables so they are packed together into one class which is parent to the others. Sign may take either the value + or −. The value of the instance variable Magnitude is selected from the list of labels, i.e.

xb	extrabig
b	big
m	medium
s	small
xs	extrasmall

with the respective sign.

The specific work of the qualitative simulator consists of mapping such defined magnitudes into quantitative ranges, which are defined by MinValue and MaxValue. The definition of these ranges for each variable is determined during the construction phase of a model, in which a user may also define an ordering between the magnitudes of different variables, i.e. variable 1 is always greater than variable 2. This ordering is inserted into a list and used for the definition of ranges when a user is not able to limit his or her variables. The single variable types are described by the classes InputVariable, StateVariable and OutputVariable, possessing supplementary attributes for their different properties. It should be noted that a class variable is used for defining the type of a variable.

Input variables may have either a constant or an increasing or an oscillating value over the whole simulation period. State variables have an initial value and an initial sign. Output variables have no supplementary properties.

Connections are defined by their predecessors and successor, and they possess a sign (indicating whether the influence is positive or negative) and a magnitude. They may be

extra strong
strong
medium
weak
extra weak

and are also translated into numerical values, which indicate how a preceding variable influences its successor. Obviously, these factors can be easily changed by a system administrator, and their suitability is checked during a test phase.

Simulation algorithm

Having defined the system, the user may choose the variables he or she wants to examine during the run as well as define the time horizon of the simulation. During the run, all variables are represented graphically, and the monitored ones have analogue gauge assigned, which indicate changes in the qualitative values (see also Chapter 6). The user can interrupt the simulation at any time, using the mouse, in order to check the situation or reset some variables and restart the simulation.

Prior to describing the computation in some detail, some remarks are needed on the ranges of variable values used. The whole range of a variable is divided by the number of possible qualitative values, actually ten. Thus, if a variable has a range from -500 to 500 and a qualitative value of small, the actual value lies between 100 and 200. In the state variables, the initial value is set to the median value of the respective range of the qualitative value, in this case 150.

In the computation, there are two different types of methods according to the variable types. In the case of output variables, the value is calculated according to

$$\mathrm{OUT}(t+1) = \mathrm{INFLUENCE}\ [\mathrm{PREDECESSORS}]\ (t)$$

and in the case of state variables

$$\mathrm{STATE}(t+1) = \mathrm{STATE}\ (t) + \mathrm{INFLUENCE}\ [\mathrm{PREDECESSORS}]\ (t).$$

The quantity INFLUENCE [PREDECESSORS] (t) is the sum of all influences of preceding variables.

The single influences are calculated with the following algorithm:

Class	**QualitativeSystem**
Superclasses	Object
Instance Variables	Name
	InputVariables
	OutputVariables
	StateVariables
	Connections

Class	**SystemVariables**
Superclasses	Object
Instance Variables	Name
	Magnitude
	Sign
	MaxValue
	MinValue
	PredecessorList
	SuccessorList
	Saturation
	Labels(−xb, −b, −m, −s, −xs, xs, s, m, b, xb)

Class	**InputVariables**
Superclasses	SystemVariables
Class Variables	Vartype Input
Instance Variables	Inform
	Parameter 1
	Parameter 2

Class	**StateVariables**
Superclasses	SystemVariables
Class Variables	Vartype State
Instance Variables	InitialMagnitude
	InitialSign

Class	**OutputVariables**
Superclasses	SystemVariables
Class Variables	Vartype Output

Class	**Connections**
Superclasses	Object
Instance Variables	Name
	Predecessor
	Successor
	Magnitude & Sign

Fig. 5.8 — Classes with their variables.†

†LOOPS shows a similar structure as KEE, presented in section 5.2. New is the notation of instance and class variables. Instance variables contain a unique value for every instantiated object, whereas class variables have the same values for every instance of a class.

1. Get the qualitative value of the preceding variable (QVPredVar).
2. Transform the qualitative value into a numeric value NVPredVar by mapping it into the respective subrange.
3. Get the qualitative value of the connection between the two variables (QVConn).
4. Transform this value into a numerical factor (NVConn).
5. Compute the numerical value of the successor (NVSucc) by the formula:

$$DELTA(SuccVar) = RANGE(SuccVar)/RANGE(PredVar) * NVPredVar * NVConn$$

$$INFLUENCE = SUM\ (DELTA)$$

6. The numerical output is retransformed into the respective qualitative value.

In addition, the derivatives of the state variables are calculated and recorded in the respective objects: see Fig. 5.9.

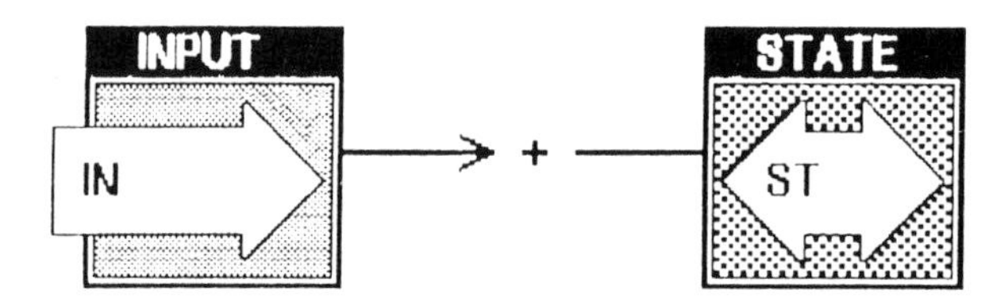

Fig. 5.9 — A state and an input variable with positive connection.

If IN has a constant value, then in the temporal evolution the value of STATE should steadily increase. But if IN and the connection is too small, STATE will not leave its original range. Taking the derivatives into account, one should remember that STATE is moving towards a range border and that its qualitative value should change. Using the following formula, one can keep track of the changes and add the value NV to the value of the variable.

$$NV = (nplus * MAX + nminus * min) / n$$

where

nplus	= number of times positive derivatives have been calculated
nminus	= number of negative derivatives
n	= nplus + nminus
MAX	= upper limit of the actual range
min	= lower limit of the actual range

In the actual version of the system only those derivatives which are significantly different from 0 are recorded, which prevents any change in the qualitative value in cases where the derivatives are too small.

The usual problem of ambiguity in qualitative systems (e.g. if a value is close to 0 or a range limit, does the qualitative value change or not?) is solved in this system by interacting with the user. In ambiguous cases, in fact, the simulator chooses the first alternative out of the set of possible alternatives. If the user, who controls the simulation by watching the different icons and their changes in time, does not agree with the ongoing calculation, he or she can interrupt, and the system automatically backtracks to the point where the first alternative was chosen. The simulation restarts at that point with the old values, but taking another alternative. Thus a tree of possible simulation runs is constructed, and a user can traverse it, seeing all the alternatives.

REFERENCES

Bezivin, J. (1987) Some experiments in object oriented simulation. *OOPSLA Proc. 87*, ACM, 394–405.

Birtwistle, G. (ed.) (1985) *Artificial Intelligence, Graphics and Simulation.* Society for Computer Simulation, La Jolla, California.

Bobrow, D. G. & Winograd, T. (1977) An overview of KRL, a knowledge representation language. *Cognitive Science* **1**, No.1, 3–46.

Bonczek, R. H., Holsapple, C. W., & Whinston, A. B. (1981) A generalized decision support system using predicate calculus and network data base management. *Operations research* **29** No.2, 263–281.

Bovee, K. D. (1982) *A guide to habitat analysis using the instream flow incremental methodology.* Instream Flow Information Paper No. 12, FWS/OBS 82–26, Fish and Wildlife Service, US Dept. of Interior, Washington, DC.

Brachman, R. J. (1983) What IS-A is and isn't: an analysis of taxonomic links in semantic networks. *IEEE Computer*, 30–36.

Bundy, A. (1985) Intelligent front ends. In: Bramer, A.M. (ed.), *Research and Development in Expert Systems.* Cambridge University Press, Cambridge, 193–204.

Cleary, J., Goh, K. & Unger, B. (1985) Discrete event simulation in Prolog. In: Birtwistle, G. (ed,) *Artificial Intelligence, Graphics and Simulation.* Society for Computer Simulation, La Jolla, California.

Clocksin, W. F. & Mellish, C. (1981) *Programming in Prolog.* Springer–Verlag, Berlin.

Dahl O. J. & Nygaard K. (1966) SIMULA — an Algol based simulation language. *Comm. ACM* **9**, 671–678.

De Filippis, M. (1988) *An Expert System for analyzing river ecosystems* (in Italian). Degree Dissertation, Dept. of Electronics, Politecnico di Milano, Milan.

Dolk, D. R. & Konsynski, B. R. (1984) Knowledge representation for model management systems. *IEEE Trans. SE* **10**, No. 6, 619–628.

Elam, J. J. & Henderson, J. C. (1983) Knowledge engineering concepts for decision support system design and implementation. *Information & Management* **6**, 109–114.

Fikes, R. & Kehler, T. (1985) The role of frame-based representation in reasoning. *Comm. ACM* **28**, No. 9, 904–920.

Forbus, K. D. (1984) Qualitative process theory. *Artificial Intelligence* **24**, No. 1–3, 85–168.

Genesereth, M. R. & Ginsberg, M. L. (1985) Logic programming. *Comm. ACM* **28**, No. 9, 933–941.

Gergely, T., & Futo', I. (1983) A logical approach to simulation (TS-Prolog). In: Wedde H. (ed.), *Adequate Modelling of Systems.* Springer–Verlag, Berlin, pp. 25–46.

Goldberg, A. & Robson, D. (1983) *Smalltalk-80: The language and its implementation.* Addison–Wesley, Reading, Massachusetts.

Guariso, G., Hitz, M. & Werthner, H. (1987) An intelligent Simulation Model Generator. Technical Report, Politecnico di Milano 87–053, 15 pp.

Guariso, G., Hitz, M. & Werthner, H. (1988) A knowledge based simulation environment for fast prototyping. In: Huntsinger, R. C., Karplus, W. J., Kerckhoffs, E. J. & Vansteenkiste G. C. (eds) Simulation environments., *Proc. European Simulation Multiconference 1988, Nice, France, June 1988*, SCS, pp. 187–192.

Guariso, G. (1988) Expert systems in water resources. *Proc. IFAC Symp. on Systems Analysis Applied to Management of Water and Land Resources, Rabat, Morocco*, pp. 123–131.

Hayes, P. J. (1979) The logic of frames. In: Metzing, D. (ed.), *Frame Conceptions and Text Understanding.* Walter de Gruyter, Berlin, pp. 46–61.

Hendrix, G. G. (1979) Encoding knowledge in partitioned networks. In: Finder, N. V. (ed.), *Associative Networks: Representation and Use of Knowledge in Computers.* Academic Press, New York, pp. 51–92.

Hogger, C. J. (1984) *Introduction to Logic Programming.* Academic Press, New York.

Kerckhoffs, E., Vansteenkiste, G. & Zeigler, B. (eds) (1986) AI applied to simulation. *Simulation Series* **18**, No. 1.

Kerckhoffs, E. J. H. & Vansteenkiste, G. C. (1986) The impact of advanced information processing on simulation — an illustrative overview. *Simulation* **46** No. 1, 17–26.

Klahr, P. & Faught, W. S. (1980) Knowledge-based simulation. In: *Proc. First Annual National Conf. on Artificial Intelligence.* AAAI Stanford, pp. 181–183.

Kulikowski, C. (ed.) (1987) AI, expert systems and languages in modelling and simulation. *IMACS Int. Symposium, Barcelona 2.6.87, IMACS Preprints.* Int. Ass. Math. Com. Simulation, Barcelona, Spain.

Luker, P. A. & Adelsberger, H. H. (eds) (1986) Intelligent simulation environments. *Simulation Series* **17**, No. 1.

Mathewson, S. C. (1975) Simulation program generators. *Simulation* **23**, No. 6, 181–189.

McCarthy, J. (1987) Generality in Artificial Intelligence. Turing Award Lecture. *Comm. ACM* **30**, No. 12, 1030–1035.

McDermott, D. (1982) A temporal logic for reasoning about processes and plans. *Cognitive Science* **6**, 101–155.

McRoberts, M., Fox, M. & Husain, N. (1985) Generating model abstraction scenarios in KBS. In: Birtwistle, G. (ed.), *Artificial Intelligence, Graphics and Simulation*. Society for Computer Simulation, La Jolla, California.

Minsky, M. (1981) A framework for representing knowledge. In: Haugeland, J. (ed.), *Mind Design*. MIT Press, Cambridge, Massachusetts, pp. 95–128.

Newell, A. & Simon, H. A. (1963) GPS, a program that simulates human thoughts. In: Feigenbaum, E. A. & Feldman, J. (eds), *Computers and Thoughts*. McGraw Hill, New York, pp. 279–293.

Nilsson, N. J. (1980) *Principles of Artificial Intelligence*. Tioga Publishing, Palo Alto, California.

Ören, T. I. (1984) GEST — A modelling and simulation language based on system theoretic concepts. In: Ören, T. I., Zeigler, B. P. & Elzas, M. S. (eds), *Simulation and Model-Based Methodologies: An Integrative View*. Springer–Verlag, Berlin, pp. 281–336.

Quillian, R. (1968) Semantic memory. In: Minsky, M.(ed.) *Semantic Information Processing*. MIT Press, Cambridge, Massachusetts.

Rajagopalan, R. (1986) Qualitative modelling and simulation: a survey. In: Kerckhoffs, E., Vansteenkiste, G. & Zeigler, B. (eds), *AI Applied to Simulation. Simulation Series* **18**, No. 1, 9–27.

Shapiro, E. Y. & Takeuchl, A. (1983) Object oriented programming in Concurrent Prolog. *Technical Report CS83-08*, Rehovot, Israel.

Stefik, M., Bobrow, D. G. & Kahn K. (1986) Integrating access-oriented programming into a multiparadigm environment. *IEEE Software, January*, 10–18.

Turban, E. & Watkins, P.R. (1986) Integrating expert systems and decision support systems. *MIS Quarterly* **6** No. 8, 121–136.

Turing, A. (1950) Computing Machines and Intelligence. *Mind* **LIX**, No. 236.

von Neumann, J. (1966) *Theory of Self-reproducing Automata*. Univ. of Illinois Press, Urbana, Illinois.

Winograd, T. (1975) Frame representations and the declarative/procedural controversy. In: Bobrow, D. G. & Collins, A. M. (eds), *Representation and Understanding: Studies in Cognitive Science*. Academic Press, New York, pp. 185–210.

Zeigler, B. P. & De Wael, L. (1986) Towards a knowledge-based implementation of multifacetted modelling methodology. In: Kerckhoffs, E., Vansteenkiste, G. & Zeigler, B. P. (eds), *AI Applied to Simulation. Simulation Series* **18**, No. 1, 42–51.

6

The user interface

The part of the software dedicated to communication with the user may take half or even more of the entire code in modern interactive computer applications. This fact underlines the importance and the complexity of interface design and construction. The interface constitutes the only part which presents the computer and its software to the user. The acceptance and success of a computer application depends heavily on this component. This chapter reviews the tasks and scopes of the dialogue module, and describes a more general framework to evaluate and to design an interface. It also discusses some human perception capabilities, describes different possible user groups and their profiles, and presents goals of and different approaches to an interface design. Several interaction techniques and hardware devices are discussed and compared. Finally, two examples are presented.

Since interfaces are not unique to DSS few items that are specific to decision support systems will be discussed. But, none the less, the respect of rules of a good interface design is also crucial for the success of an EDSS.

Due to the fact that the development in this area is very dependent on progress at the hardware level, this chapter is a rather heterogeneous, but compact, listing of tools. It is nearly impossible to describe in detail all hardware trends and devices. So it has been limited to a short overview enriched by some recommendations. Specific descriptions from the different manufacturers will provide more information.

From a philosophical point of view, the use of the terms 'communication' and 'interface' for an interaction between a machine and a human being can be criticized. It implies that both components of a man–machine system can be viewed as free and independent, as is normally the case in dialogue between humans. However, up to now at least, this has not been the case. 'Communication' originates with a system builder or programmer, who has to foresee and to implement all possible interactions between a computer and its future users and who determines the form and frequency of interaction. Moreover, a software designer puts some restriction on possible user groups by choosing specific interaction styles, thus favouring one group and injuring another. Rather than a form of interaction between humans and machines, such communications take the form of strongly hierarchic interactions between humans. While this view (Nake 1984) is shared by the authors, the term 'communication' for man–machine interactions will be used as it is widely understood in the area of computer science.

6.1 HUMANS AND MACHINES

Primary design goals, such as proper functionality, reliability, security and integrity of implemented software, have to be guaranteed by an interface module. The interaction with a user constitutes a danger to the correct functioning of software, since it is nearly impossible to foresee all possible user actions. Historically, interface design stressed the controlling scopes, describing human users as external factors whose possible interferences had to be properly constrained. With the spread of computer products for personal use and the growth of the user community, more importance had to be given to the users and their acceptance of systems. With the steady increase in the use of computers in normal commercial environments, in industry, in offices, and in family homes, the importance of the human factor has grown and has been recognized. Typical users of interactive systems, such as EDSS, are no longer experts in computer science. Thus, an interface not only has to guarantee error-free use in order to avoid system breaks, it also has to support the user, give him or her advice in critical situations, and explain the underlying logical model of the application. More important, the interface has to support the feeling that the computer serves the user and that it is under his or her control.

Table 6.1 gives a tentative distinction between capabilities of a human and a machine. As can be seen, the 'more intelligent' features remain on the human side. Machines perform repetitive and complex calculation tasks better, whereas humans have an ability to justify and to decide in unforeseen situations, a talent a machine cannot achieve. Humans play (and they will continue to play) the most important and relevant part in man–machine systems.

These differences have to be taken into account in the design of a man–machine system. The separation of tasks has to be such that the machines adapt to the human in order for the result to be both satisfying and successful. An interface is more than a combination of hardware devices and some software modules. Because it is the part which allows the user to interact with, to control and to determine the work of the system, it forms an integral part of the software and its design process. Thus, when system designers see that certain interfaces are not well accepted, they should no longer be used. User needs have to be taken into account at the very beginning of the design of the whole system. The scope of the dialogue component is to bring an application closer to its human users and to adapt it to the needs and abilities of humans.

Table 6.1 — Capabilities of humans and machines (according to Shneiderman 1987)

Humans usually do better	Machines usually do better
Sense low level stimuli	Sense stimuli outside human's range
Detect stimuli in noisy background	Measure physical quantities
Recognize constant patterns in varying situations	Store quantities of code accurately
Sense new events	Monitor prespecified events
	Make rapid responses
Remember principles	Recall quantities rapidly
Retrieve pertinent details without *a priori* connection	Process quantitative data in prespecified ways
Draw upon experience and adapt decisions to situation	
Select alternatives if original approach fails	
Reason inductively: generalize from observations	
Act in unforeseen situations	Perform repetitive commands
	Exert highly controlled physical force
Apply principles to varied problems	
Make subjective evaluations	
Develop new solutions	
Concentrate on important tasks	Perform several activities simultaneously
	Maintain operations under heavy information load
Adapt response to changes in situation	Maintain performance over long time

6.2 THE INTERFACE IN AN EDSS

Re-examining the proposed architecture of an EDSS (see Fig. 6.1), the importance of the interface is clearly reflected in its incorporation as an autonomous module. Its role is even more important in DSS applications since typical users will not be experts in computer science and not even be very frequent users.

As shown in Fig. 6.1, the interface constitutes a unique opportunity to utilize the various possibilities of the system. The user can call the different bases without leaving the system and can use the system to pass information from one step of work to another. The ability of the interface module to keep all these different bases inside the system shows its integration facility but

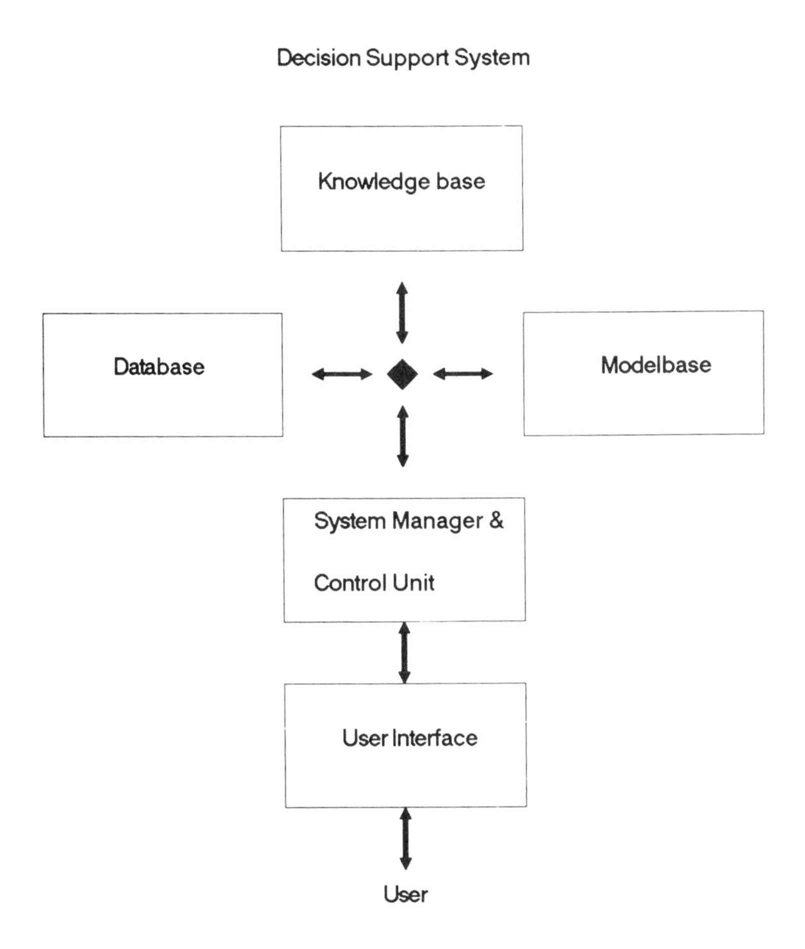

Fig. 6.1 — The extended architecture of an EDSS.

also its capability in supporting many possible actions. This ability to present the system in a unique way is valid not only for the software; there should also be support for several hardware devices which should be accommodated to specific tasks and human abilities.

The design as a distinct module with a well-defined connection to the rest of the system has one more advantage: the possibility of a step-wise modification of the system. Changes in the system are possible without changes in the interface and vice versa. The incorporation of new hardware devices does not require any change in the logic of the software, and the adaptation of the interface to new users does not call for a modification of the entire software. A general framework of the interface and its design is presented in the following sections.

6.3 HUMAN PERCEPTIVE AND COGNITIVE ABILITIES

Human perceptive abilities and limits of those abilities form the basis of proper interface design. Such human limits must be respected in order to avoid severe physical problems and to establish a satisfying working environment. They may be characterized by the list in Table 6.2.

Table 6.2 — Human perception abilities concerned with DSS interfaces

Spectral range and sensitivity
Colour vision and deficiencies
Peripheral vision
Flicker sensitivity
Contrast sensitivity
Low-light vision and adaptation times
Bright light vision
Motion sensitivity
Legibility of text
Response time to varying visual stimuli
Capacity to identify object in context
3-D vision and depth perception
Viewing distance and angle

Taking as an example the legibility of text, an illustration will show how human limits should be taken into account. The legibility of written text can be increased if the letters are not all capitals. Tinker (1965) describes an experiment where the speed of reading was slowed down by 14% to 20% if only capitals were used. Obviously humans do not read letter by letter as in a spelling process; it seems that they read whole words at once and also recognize words by their shape. In the following the right-hand phrase should be easier to read than the left-hand one.

THIS IS AN EXAMPLE This is an Example

In Tinker (1965) more advice is given on improving the legibility of text: the depth of text should range from 9 to 12 points (1 point is 1/72 inch or 0.35 mm vertically); the line length should be from 2.3 in (5.8 cm) to 5.2 in (13.2 cm); and the spacing between two lines should be approximately 2/72 in (0.7 mm). In Kolers *et al.* (1981), it is shown that on computer screens such as CRT's (cathode ray tubes), the reading speed with 40 characters per line is 17% slower than with 80 characters. Another experiment is presented by Radl (1980), which concludes that a negative contrast on a screen (dark letters on light background) produces better results in reading than a positive contrast, if the refresh rate of the screen is sufficiently high.

A test to determine the best speed for reading text on a screen was conducted at display rates of 150, 300, 1200 and 9600 baud (bits per second, which correspond to 15, 30, 133 and 860 characters per second); the best result was at a speed of 30 characters per second (Tombaugh 1985). Readers are able to keep up with the display at lower rates; if the speed is too fast, comprehension and satisfaction deteriorate. On the other hand, if screens can be filled instantly, users do not seem to feel stress and work well.

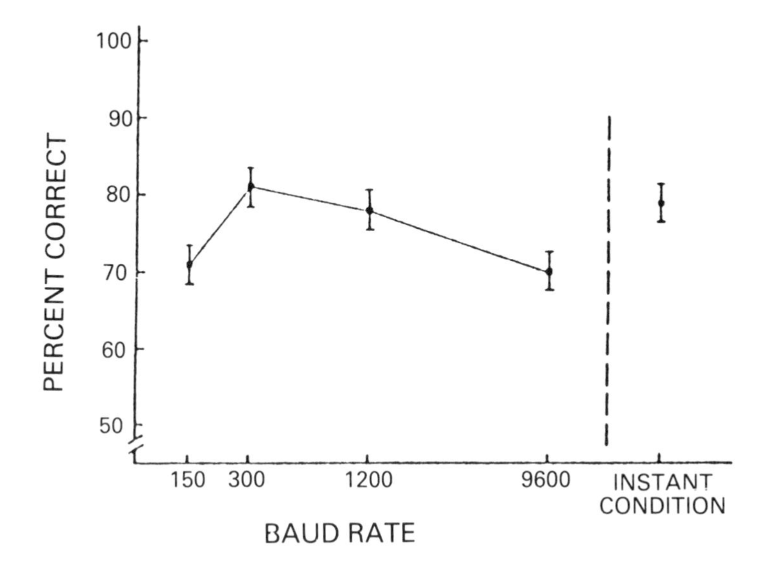

Fig. 6.2 — Comprehension scores according to Tombaugh (1985).

With regard to human cognitive abilities, i.e. the ability to understand and evaluate a signal, a distinction can be drawn between short-term memory (STM) and long-term memory (LTM). Memory is not a one-level system; it is believed to be a complex system of several levels of storage with a central large-capacity store (LTM). STM seems to depend on some other stores, such as input and output buffers (Thomson 1984). As an example of the two different stores, one might first consider telephone numbers, which are retained in the memory for a short period, and second, basic engineering principles, which are memorized and retained for a long time at a central place. According to Miller (1956), approximately seven units or 'chunks' of information can be kept in the memory for about 15–30 seconds, depending on personal acquaintance with a problem area. The short-term memory, with its integrated input and output buffers and the exchange of information between them, shows similarities to a computer. In order for information to be retained in the STM, however, it is important that no disturbance occurs between the memorization and the recall of information.

Long-term memory is the least understood and most complex. It seems that information is stored in the form of a semantic code on a very high level associated with a subjective judgement. This means that people are able to memorize the meaning of information, but have problems in keeping up with all the details. Memorizing of information in the LTM, i.e. the transfer from the STM, takes a rather long time. The retrieval process seems to work in an associative way, involving complex operations such as finding similarities or recalling structural equalities in the information.

Response times of computers, as defined in Fig. 6.3, should be set

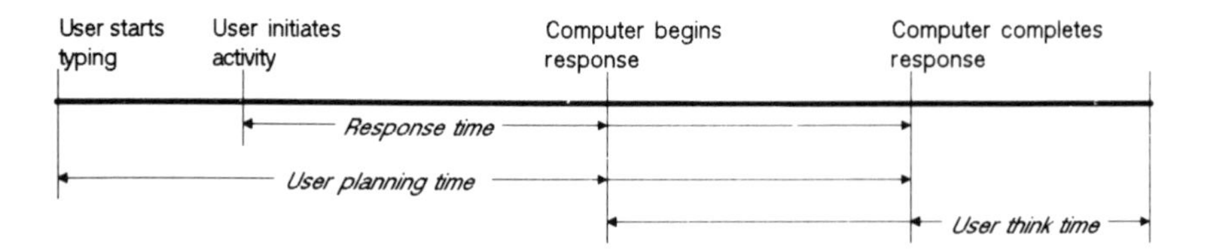

Fig. 6.3 — Definition of response time.

according to the STM capabilities, i.e. limited in time and amount, to enable fast interaction. Other facts which influence the behaviour of a user may be expectation of the response time, feedback about progress of a task, or knowledge about the complexity of an executed task.

In general, it seems to be difficult to derive rules for time limits for desirable response times though Table 6.3 cited by Miller (1968), lists some acceptable delays.

6.4 USER GROUPS

A user possesses a conceptual model of the system, i.e. a mental view of the system and the logic incorporated in it. The interface design has to foresee such a model and should give a consistent representation of it to a user. It can support the transparency of the system's work without burdening the user with unnecessary technical details. The conceptual model lies at a higher level than the various implemented details; in a way, it is a reflection of the user's expectation of the system's functions. The designer, knowing the future users and their profiles, can enforce the construction of such a conceptual model. One way to derive necessary distinctions in human behaviour can be to review different knowledge areas and levels when using the system.

Fig. 6.4 distinguishes between knowledge about the computer and task-specific areas. Double boxes indicate primary information sources, the other boxes secondary ones. Recalling secondary information is necessary for inferring primary sources which are lacking. Primary knowledge that is missing, for example in the physical use of an interface, may be repaired by existing knowledge about other machines. On the other hand, using pre-

Table 6.3 — Response time classification

Classification	Acceptable time delay (seconds)	Comments
Echo characteristics	<0.1	Example is visual feedback from screen
Conversational requests	<2	Most editing tasks
Searches:		
String search	<4	Two seconds would be preferable, up to 4 are acceptable
Browsing	<0.5	
Task completion:		
Program execution	<5	Depends on program, delays should not be too long (messages)
Log on or log off	<15	
Recovery from system failure	<15	If it takes longer, the system should inform the user

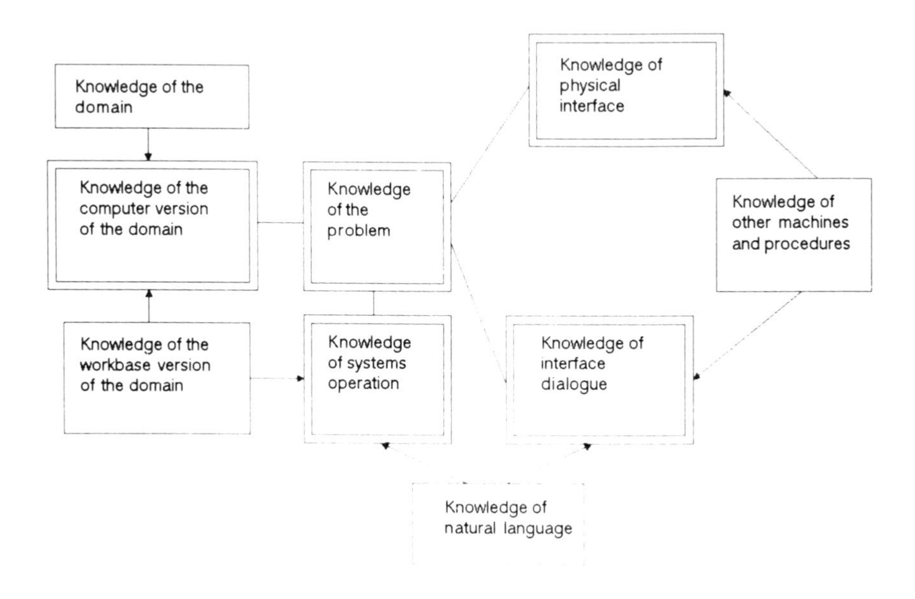

Fig. 6.4 — Knowledge in interaction with a computer (Hammond and Barnard 1984).

existing knowledge in the absence of primary knowledge could result in misleading actions, for example the improper placement of the command END in a sequence of commands. This may make the user think that END indicates a real end of all tasks, whereas it might only indicate an exit from an intermediate routine.

The following classification of user communities and their correlated interface techniques can be drawn up with reference to Fig. 6.4:

— Novice user: They have little knowledge about the computer system and possibly also little knowledge about the problem domain. All boxes, with the exception of the knowledge of natural language, are weak. Such users may show anxiety in front of the machine; they do not know what will happen and how they might be able to solve problems. In this case, good tutorials, menus with clear terms, and positive and exhaustive error messages may be of help and form a good strategy.
— Knowledgeable intermittent users: They know something about computers and the general problem area, but little about the specific implementation. They probably constitute the biggest group. The double boxes are weak, with the exception of the knowledge about the problem. For these users, the recognition of their conceptual model of the problem and its structure is of importance; this might be done by the user by recalling from his or her memory similar implementation on other machines or former uses of the actual system. Useful techniques might be online help screens and the maintenance of consistency of actions to support a clear model of the system.
— Frequent users: They have knowledge in nearly all areas of actual implementation; in particular, their knowledge about possible actions and responses is well developed. They may lack some general knowledge about the problem domain and also about computers. Sometimes online semantic explanation windows might be of help; moreover, short response times, a command language and a possibility of abbreviations are of importance.

An interface construction which satisfies the needs of all groups seems to be difficult; the scope depends on the foreseen area and user groups. In EDSS, groups 2 and 3 will probably dominate. Facilities such as help screens on demand, the possibility of limiting the amount of messages, and the ability to switch between menu and command-driven control might be useful.

Dimension of a user's behaviour

The different dimensions of a user's behaviour also constitute measurable goals in the design process. These human factors are important for setting priorities and for evaluation purposes (Moran 1981):

— Functionality: describes the range of work which a user can fulfill with the system,

— Learning: corresponds to the time it takes a human to learn to do a given task,
— Time: is the time it takes to perform a specific task in the system,
— Errors: describe the number and severity of errors made by fulfilling a specific task,
— Quality: is a measure of how a given task is fulfilled,
— Acceptability: corresponds to the subjective rating of users,
— Retention: measures the time a user is able to maintain his or her knowledge, and
— Robustness: describes how well a user can adapt his or her knowledge to new tasks.

It is nearly impossible to fulfil all these goals at the same time. There are always trade-offs in any design choice. For example, the wish to shorten the time to fulfil a specific task conflicts with the goal of minimizing the number of errors made during the work.

6.5 INTERFACE DESIGN APPROACHES

The overview of perception and cognitive abilities, user profiles and design goals in the preceding sections should provide a framework for designing an interface. This design process is an application of the understanding or 'psychology' of a user. The design has to foresee and to incorporate possible user reactions and behaviour. This cannot be done purely by measuring user acceptance; there should be a theory about or a model of the user for explaining and foreseeing his or her behaviour. Thus, the spectrum of different approaches to designing an interface ranges from rather empirical approaches to strongly formalized models. The empirical approaches are used for the evaluation of distinct design alternatives or to test some general design features. Experiments, where users interact with prototypical installations, are used to evaluate their performance (see also the previous section for different dimensions of a user's behaviour).

Theoretical approaches try to explain the user's behaviour with the help of abstract models, differing in their degree of abstraction and their possibilities for direct application to the design process. In the so-called calculational approach, as a first example of a theoretical model, user behaviour is described with the help of a model which allows for predictions in the sense of calculation. A psychological theory is built in the form of an explicit information-processing model. It describes the single mental and physical steps a user has to make to perform specific operations. Once encoded in the form of a model, the behaviour can be calculated. An example is the Keystroke–Level model (Card *et al.* 1980) for explaining and calculating times (i.e. for finding a key, for striking, for thinking about actions, for moving the hand) applied to a text editor. With this model, the time it takes to a specific action can be calculated in advance.

Another model is the syntactic/semantic one (Shneiderman 1987), which explains the user's knowledge on different levels and in different domains.

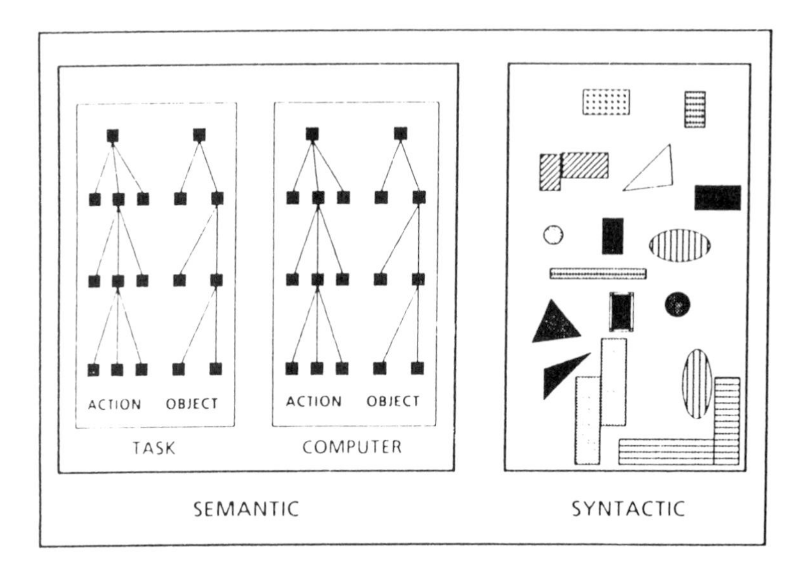

Fig. 6.5 — Syntactic/semantic model according to Shneiderman (1987).

Syntactic knowledge is extremely device and system dependent and can be easily forgotten. It concerns information ranging from the proper use of input or output devices to the correct syntax of specific interaction commands. Semantic knowledge, on the other hand, can be divided into computer and problem (task) specific domains. In addition, it can be separated into objects and actions which can act on objects. The model reflects a hierarchical structure. A semantic object, for example in the domain of the computer, might be correlated with the concept of storage. At the highest level, there is an object which describes the concept of stored information; at the middle level, there are directories and files; and at the lowest level there are segments, records, or lines. Correlated with these objects are actions that have to do with the storing and retrieving of information. There are certain actions at the different levels, such as the creation or storage of files, the assigning of disk drives, and the giving of access permissions, or, at the lowest level, specific storage and retrieval commands. The hierarchical structure of this model serves well for learning and reusing concepts. Once the general process of information storage is well understood, it is easy to recall the corresponding low-level actions. A similar hierarchy and division can be found in the task-dependent area.

Another formal approach to designing an interface is the use of state transition diagrams. They define and also limit the set of possible actions of a user, depending on the actual state of the whole system (Wasserman *et al.* 1986). Such a diagram can be described as a finite automaton (see Fig. 6.6). It gives a functional description of possible actions, and thus is well suited to guaranteeing a correctly working system, but does not give attention to the user's psychology.

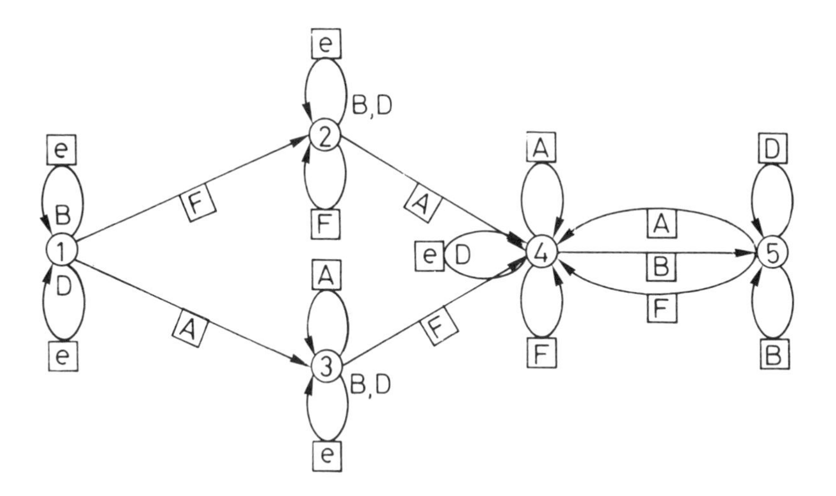

State	Input F	A	B	D
1	2	3	1 e	1 e
2	2	4	2 e	2 e
3	4	3	3 e	3 e
4	4	4	5	4 e
5	4	4	5	5

Fig. 6.6 — A simple state transition diagram with related state-input table, where e indicates error.

Still another approach to designing a correct user interface is described by Chi (1985). He formalizes the interaction with a mathematical algebra of axioms of possible actions, which defines correct and permissible operations on consistent states of the system to pass to another consistent state.

The design of an interface has to be seen as a process, and therefore it cannot be represented statically. Most probably, new goals will be derived during this process. Designing is dynamic in nature. A possible way to involve the future users in the formulation of goals and to incorporate them into the process might be the so-called participatory approaches. Another possibility is represented by prototyping systems. One can generate masks as a representation of the interface without the underlying software procedures and give them to user groups for testing and evaluation (in close relationship to the empirical approaches). Goals, which could be tested in confronting future users with such a prototype, might be functionality, acceptability or other factors from the list of user behaviour. The choice of possible subjects and of the test environment is of importance.† For example, if the interface is to be used by novice users in a noisy environment, these conditions should also be fulfilled in the testing circumstances.

Design rules

As an intermediate result of the framework developed so far, some general design rules (following Shneiderman 1987) are listed below. They have to be seen more as overall guidelines, while exact adherence to them depends on the distinct application:

- Consistency: guarantees a consistent sequence of actions and rules in similar situations. It also provides coherent terminology during the use, i.e. in the dialogue, display and data entry.
- Enable shortcuts: frequent users should have the ability to use macros, abbreviations, and faster displays. Attention should be paid to response times.
- Minimal input: redundant data input actions of the user have to be avoided. The system itself has to remember the tasks already completed and to infer, if possible, supplementary necessary information by itself.
- Informative feedback: the system should always give feedback, the amount of information depends on the complexity of the specific task and the user's knowledge about the task. Attention should be paid to a visual presentation.
- Design to yield closures: the sequence of actions should have a beginning, a middle and an end. The system feedback should respect and reflect this sequence.
- Simple error handling: sufficient error messages and a good explanation of possible recoveries should be given. Messages should not demoralize the user.
- Easy reversal of actions: user actions should be reversible. This reduces fear since actions may be undone and alternatives are available.
- Support control of system: a user has to feel that he or she masters the system and that it responds to his or her actions. On the other hand, a user should not be faced with unnecessary technical burdens.
- Reduce short-term memory load: since the content of the short-term memory can easily be overwritten, the speed of interaction has to be adapted to its abilities. The user should be able to recall previous actions. The system should not stress the user with too complicated and overloaded screens.

† Statistical methods for evaluation will not be discussed here.

— Familiar displays: screen formats should be familiar to the user. Well-known diagrams, tables or figures should be used.
— Compatibility: an attempt should be made to remain compatible in the use of the system, for example, the same presentations should be used for data display and data entry.

6.6 DIALOGUE TECHNIQUES

In the following section some of the basic techniques and tools used to create an interface will be described, based on the following main interaction styles:

menu
form fill-in
command languages
direct manipulation
natural language interface

Menu

Menus present the user with a list of well-distinguished alternatives for the

```
                    MAIN MENU
                    ---------

          1. WHAT IS THE SCOPE
          2. WHAT DO I LEARN
          3. WHAT DO I NEED TO UNDERSTAND IT
          5. WHAT ARE THE SINGLE CHAPTERS
          4. HOW CAN I USE IT


     SELECT A NUMBER AND PRESS CARRIAGE RETURN
```

Fig. 6.7 — A prototypical menu with five alternatives.

continuation of the program's work. The user selects one of the alternatives, or, in the case of multiple selection menus, may choose more than one item. The single selectable points may have associated numbers or letters for proper selection from the keyboard, or selection may be via pointing devices such as special keys, mice, lightpens or touch screens. There are several forms of menus, such as those in the Fig. 6.8.

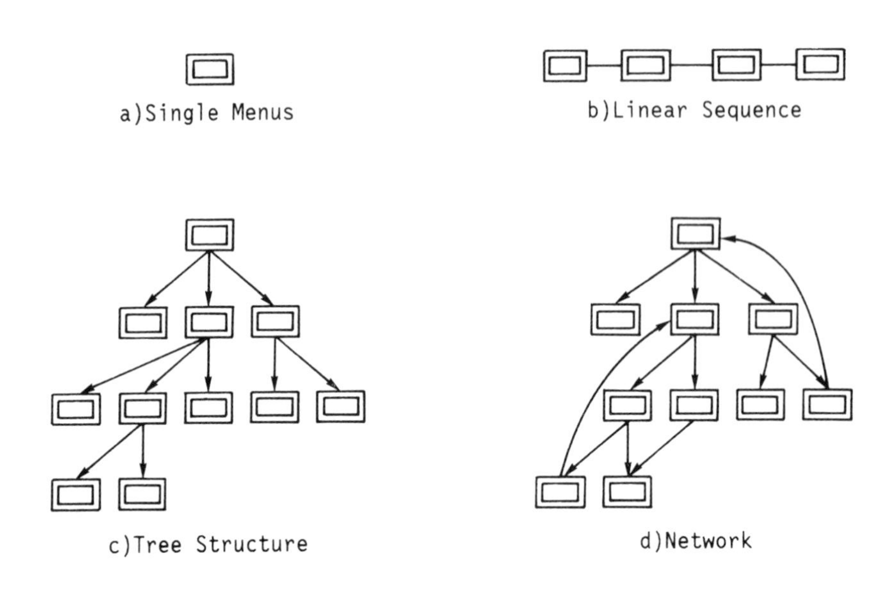

Fig. 6.8 — Types of menus: (a) single, (b) sequence, (c) tree, and (d) network.

A single menu collects all possible alternatives at one level, whereas in a sequenced form a series of alternative items have to be chosen. Common examples are the questions for setting up a printing device in some word processors. In bigger and more complex applications, trees or networks of menus are used. In the latter cases, suitable trade-offs in the choice of depth and breadth of these structures have to be found. In general, it is better to group the alternatives in fewer levels, which allows for easier orientation in the hierarchy (see Kiger 1984). When the structure of the menu tree becomes complex, positive features for orientation are necessary to indicate the actual position in the tree and to provide additional menu options, such as to pass immediately to the root level or to view the other menus of the same level.

The semantic structure of the single groups is also of importance. Moreover, attention should be paid to naming the single items in a clear and meaningful way (see, for example, Glasauer *et al.* 1985). Menus are a proper interaction style for novice users. If this interface style can be provided with a simple mechanism to pass between the single menu tree levels, it allows simple orientation and can also give the user a consistent model of the contents of the program being used.

Form fill-in

Fill-in forms are well suited for guided data entry and allow in parallel the display of information. This provides a context-sensitive way of asking the user for input data, depending on the actual state of the calculation.

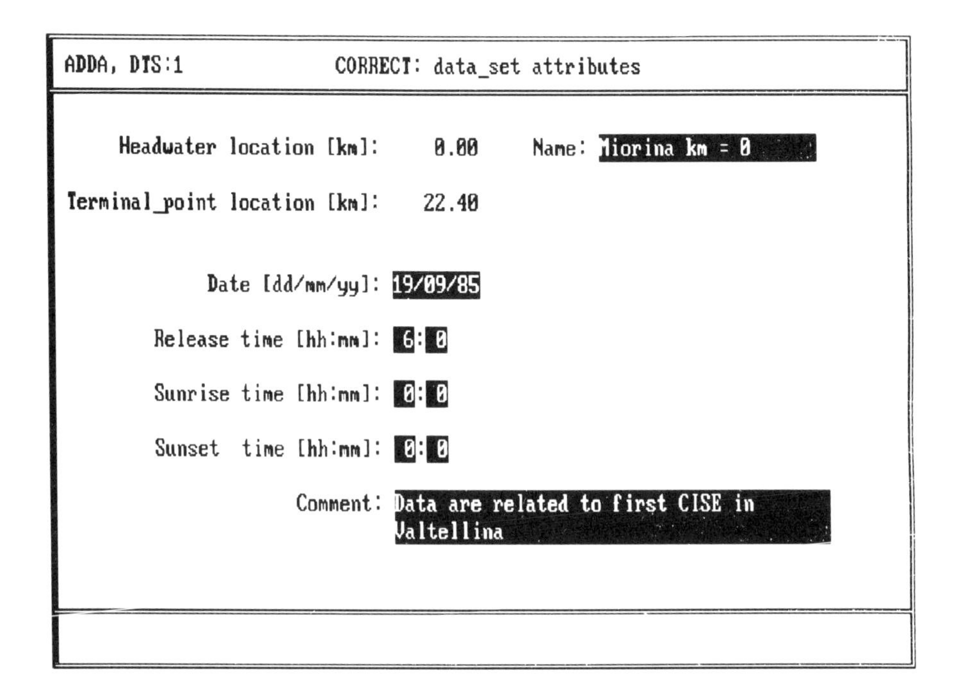

Fig. 6.9 — Form fill-in example (from WODA, see Chapter 4).

The main input device is the keyboard; special keys or a mouse may also provide the possibility of moving from one input field to another. Typical applications can be found in many commercial software products.

The form fill-in is also a good tool for novice users, but may also be used to advantage by experienced users. Well-chosen field names and powerful possibilities for correcting erroneous inputs seem to be the key feature in guaranteeing easy acceptance of this technique. Another positive feature is the limitation on possible inputs in specific fields, for example by allowing only digits in date fields.

Command language

This style represents the oldest and most commonly used form of interaction. Starting from early forms of codes that were nearly impossible to understand, command languages have passed through a long evolution process to much more comprehensible forms, using mnemonic methods. This mode of communicating with the computer is preferred by experts, since it allows fast and compact formalization of instructions. Common examples are operating systems on nearly all machines. The syntactic construction is nearly the same as in most languages: commands are followed by arguments and parameters, which specify what and how something has to be done with specified objects. The command

COPY File1 File2

is an example of the MS–DOS operating system for PCs, which tells the system to copy the contents of File1 to a file which is named File2. In the construction of command languages, attention should be given to the ordering of arguments, and, if possible, the language should express structural features. For example, it is useful to group all instructions handling storage operations around one central command. Using meaningful names for commands is surely appreciated, though some experts seem to prefer rather strange constructions. Ways of speeding up the interaction can include, for example, the assignment of special or function keys to the most commonly used commands, and the use of short codes for commands.

Direct manipulation

In some interactive systems, direct manipulation of program objects is possible. The user sees the immediate results of his or her actions, which are displayed on the screen. Examples of such interaction styles are used in desktop publishing programs, such as the Ventura system in Fig. 6.10. They are of the type WYSIWYG (What You See Is What You Get), thus showing the user the document in the format in which it will be printed later on a sheet of paper.

Advantages of this dialogue technique are the visibility of objects, the immediate presentation of the results of actions, and the ease of reversing actions. Thus, a user can try different strategies without having to fear that his or her actions might destroy some portion of the work. Moreover, this approach allows a complex command syntax to be replaced by simple actions executed on the screen. Other examples of such interaction styles are full screen *text editors*, such as WORD or CHIWRITER, or *spread sheet* programs, such as LOTUS.

Still another use of this method is constituted by interactive video games or interactive educational programs. For example, the programs described in Brighetti *et al.* (1987), which were developed for teaching ecological principles in a new and animated way in elementary schools in Italy, have shown very good results.

In the program of Fig. 6.11, the pupil moves the trout with the four direction keys and catches his or her food by using the RETURN key on the keyboard. His or her actions, successes and failures are memorized and used for the estimation of the parameters in the differential equation, which describes the increase or decrease of trout numbers in an ecosystem.

As a final example of direct manipulation, a QBE (Query By Example) database query system is presented in Fig. 6.12. It was also developed for novice users.

The user fills in the values of the single attributes of the relation shown,

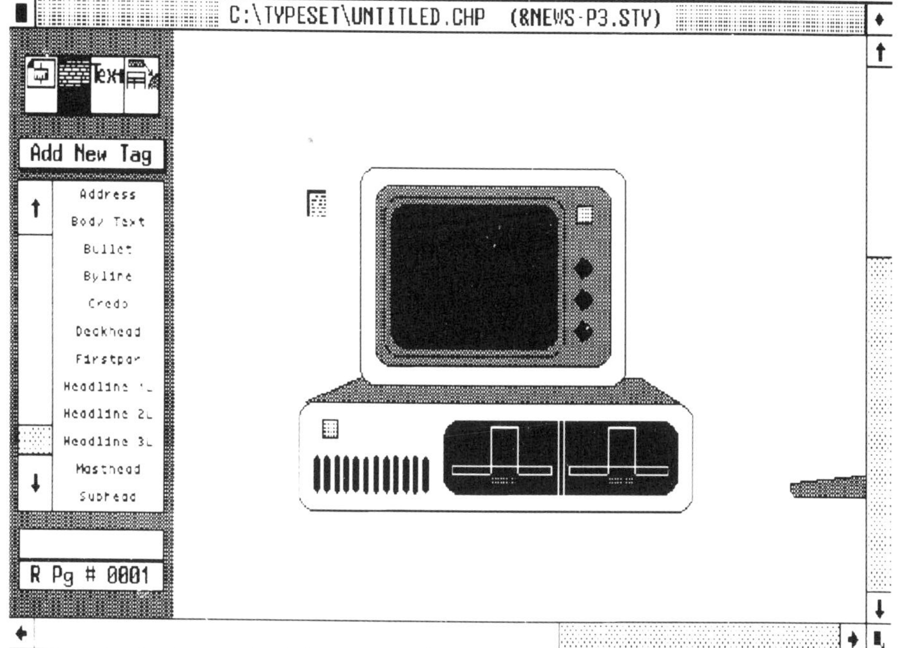

Fig. 6.10 — A typical screen of the desktop publishing program Ventura Publisher.[†]
[†]Ventura is a registered trademark of Ventura Software, Inc.

and the system prints the tuples which correspond to the attribute values on the screen. This example might also be seen as an application of fill-in forms; but the fast access to the data and the possibility of working directly on these data are a demonstration of direct manipulation methods.

Direct manipulation is well suited for novice users, once they are acquainted with the necessary hardware device for controlling the input. But it may also be appreciated by experts. On the other hand, there might also be some problems with this approach, such as:

— Sometimes it might be hard to identify the meaning of a designed object.
— Not all semantic concepts have graphic representations that are easy to understand.
— Direct manipulation, which heavily relies on graphics, also depends on the resolution of the screen and the space available on the screen.
— Experienced users may prefer the keyboard together with a command language for a faster input.

Natural language interface

Natural language interface (NLI) is one of the main research areas in AI, but only a short discussion of the scope and limits of this approach will be given here. Two main areas can be distinguished: understanding of freely typed

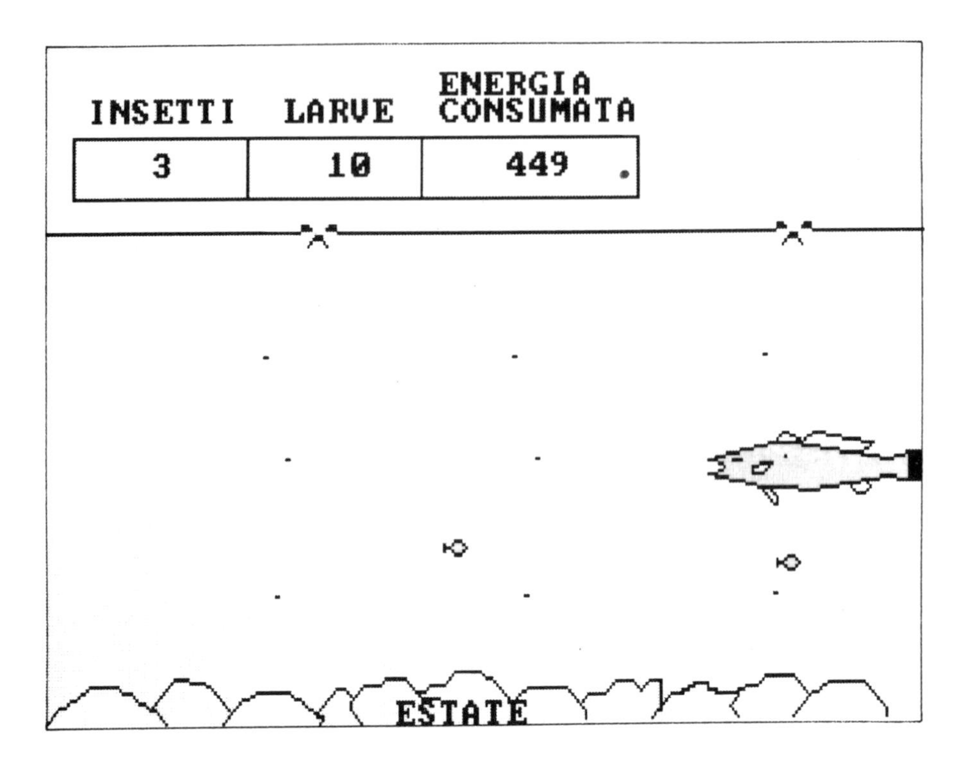

Fig. 6.11 — A program to teach ecology showing a trout in search of its food.

sentences in natural language and speech processing. Technically, in both areas, first the input has to pass a syntactic parsing and secondly a semantic interpretation has to be found.

In a narrower sense, NLI is restricted to the first area. Normally, the typed-in sentences are translated into some form of command language. It should allow the user free communication with a computer that is not limited by syntactical rules. But, in actual systems, some severe limits are put on the users. Since a system cannot understand all the background and history of a user, which would be necessary for the comprehension of all sentences, the user is modelled, i.e. some formalization of his or her knowledge, habits, etc. is predefined and embedded in some previously known context. Still another problem is constituted by the usable notation. At first glance, the notation seems to be free of restrictions. The user gets the feeling of being able to type every possible phrase. But there are always some limits to grammar and vocabulary. If the specific rules, which the user does not know, are not respected, such systems tend to ask a lot of questions. Finally, the user may become confused since he or she does not understand where these limits are. Considering these drawbacks, it may seem preferable to use a restricted interface technique, which offers a concise syntax. This enables the user to get a clear concept of the computer application and his or her possibilities. In some areas, however, NLI has been successfully applied,

PERSONAL

DATENABFRAGE SEITE 1

P#	LIS
ZUNAME	
VORNAME	
GESCHL.	M
GEB.DAT.	< 1954
WOHNORT	< > WIEN
PSTLTZ.	
STAATSB.	
FAMST.	L
# KINDER	> 1
WEHRD.	
SCHULB.	UNI
ERL. BERUF	
FREMDSPR.	ENG
KURSTAGE	
BETR.ZUG.	

CONTROL-TASTEN: ↑ ↓, Home, Delete, RETURN

PgUp = VORIGE SEITE PgDn = NÄCHSTE SEITE Ins = EINGABEN LÖSCHEN End = ENDE

Fig. 6.12 — QBE screen used for a query in an educational database application (Neunteufel 1985).

for example for document retrieval (Guida and Tasso 1983), database queries (Waltz 1978, Hendrix *et al.* 1979) and the extraction of newspaper texts within a limited context (Schank & DeJong 1979).

In *speech processing*, three main areas can be isolated: discrete word recognition, continuous word recognition and speech generation. In the first two cases, the problem of the recognition of spoken words is added to the problems mentioned in the previous paragraph. First, the sound of spoken words has to be transformed to written letters. Such a system should show advantages in cases where the human is occupied or handicapped and cannot move his or her hands. There are systems with discrete (i.e. single) word recognition of a specific person with rather limited dictionaries; the system has to be trained by its future user. Research is presently going on with larger vocabularies and without the restriction to a specific person.

Continuous speech recognition has the additional problem of the recognition of word boundaries and of different speeds and pitches of voices. So far only experimental systems exist. A final example is the generation of speech which works on two bases: the first approach uses natural speech where stored waveforms (corresponding to syllables) are concatenated to form words; in the second approach, synthetic speech generation, rules are used to generate speech synthetically. These final interaction techniques complete the overview, although it seems that real, practical application will be delayed for some time.

6.7 HARDWARE DEVICES

In this section, a brief overview of the different hardware devices for input and output will be given; a more detailed discussion of these devices is not within the scope of the present work and would always remain incomplete because of the fast progress in this area.

For historical reasons, the *keyboard* still plays the most important role in data input. Its layout goes back to 1870, with Shole's typewriter design with the QWERTY ordering of the keys. The keys are ordered to place frequently used letters far away from each other in order to avoid jamming, because at that time the technology was not well developed. Although this layout is no longer necessary, newer keyboard forms have not been successful. On computer keyboards, keys for special actions were added to this layout. Function keys, which can also be programmed by the software, were integrated mostly on the head or the side of the letter keys. Other specialized keys are used for cursor movements, to finish a data entry action or to switch the meanings of keys.

Pointing devices use the screen and a supplementary device for input. The user has to point with this supplementary device to a point or region on the screen. The coordinates of that screen position are taken as input to the program. These devices can be considered user friendly, since the user can directly point to the objects or actions of interest, and no special language has to be learned. These devices are applicable to different tasks:

— Select: choose from a set of items on the screen,
— Position: choose a point on the space on the screen, e.g. place a window,
— Orient: choose a direction, e.g. rotate a symbol or indicate the direction of an object,
— Path: perform a series of position moves and orientation operations, and
— Text: entry of text, the user pointing to a position where the text has to be placed.

These pointing devices are grouped into direct (lightpen, touchscreens) and indirect (mouse, trackball, joystick and graphics tablet).

A *lightpen* allows the user to point to something on the screen with a device similar to a pencil. Most lightpens have a button which has to be pressed to perform an input when the correct position is found on the screen. Possible problems are arm movements, which may tire the user and also hide the screen. Furthermore, these pens, which are based on a optic fibre sensor, may be heavily disturbed by ambient light. *Touchscreens* allow the user to point directly with his or her fingers on the screen, and thus the user does not even have to learn to use a lightpen. Such screens perform very well when the users are really novice, but the problem of obscuring the screen remains. In addition, if the objects to be chosen are small, there might be a problem of accuracy; and if the software accepts the touch immediately, it might lead to incorrect inputs if the screen is touched unintentionally.

Indirect pointing devices are physical instruments which move the cursor, i.e. the actual and specially signed data entry point, on the screen. There is no problem of hand movements, since these devices can be placed close to the keyboard, which is still in use in the case of data entry, but there are more problems of hand–eye coordination than with the use of direct devices. The most recommended of these devices is the *mouse*. Here, the hand remains in a comfortable position near the keyboard. A mouse allows the cursor to be positioned very precisely. It also has buttons on top to activate different operations once a position is located.

A *trackball* may be considered as a converted form of mouse and is implemented as a rotating ball that moves the cursor on screen. There are also additional buttons which allow for activation of a command. *Joysticks* are appealing for following objects on the screen and are mostly used in interactive video games. A *graphics tablet* contains a touch-sensitive flat surface which is separated from the screen and a supplementary device such as a pencil. The tablet comes into contact with this second device, and transmits the respective coordinates to the computer. Tablets may be of several different sizes and are useful for graphic inputs, which are necessary for example in geographical information systems and computer aided design (CAD). Because the available choices are printed on the screen, tablets also allow a novice user to work successfully.

A comparison between the different devices with respect to positioning, speed and accuracy is given in the Fig. 6.13.

It follows that direct devices are faster but less accurate. In applications where objects are large and the users are mainly novices, they are preferable. On the other hand, if a keyboard is needed, i.e. for normal data entry by the user, a device such as a mouse can be used to advantage, since it is rather fast and precise and the hand can remain close to the keyboard.

Displays are now the primary devices for giving feedback to the user, rather than, as in former times, a printing console. Screens are made in various sizes with different powers of resolution in both monochrome and colour versions. Monochrome displays may work on the basis of a cathode ray tube (CRT), which is similar to a television device with an electron beam. Negative and positive contrast (i.e. difference of background and foreground colours) is possible. Of main importance for the human eye is the refresh rate of the screen. Other technologies are plasma panels (limited resolution, but flicker-free) and liquid crystal displays (LCD), as used nowadays in many portable computers. Colour displays use RGB (Red Green Blue) for producing all other colours. The resolution is usually not as

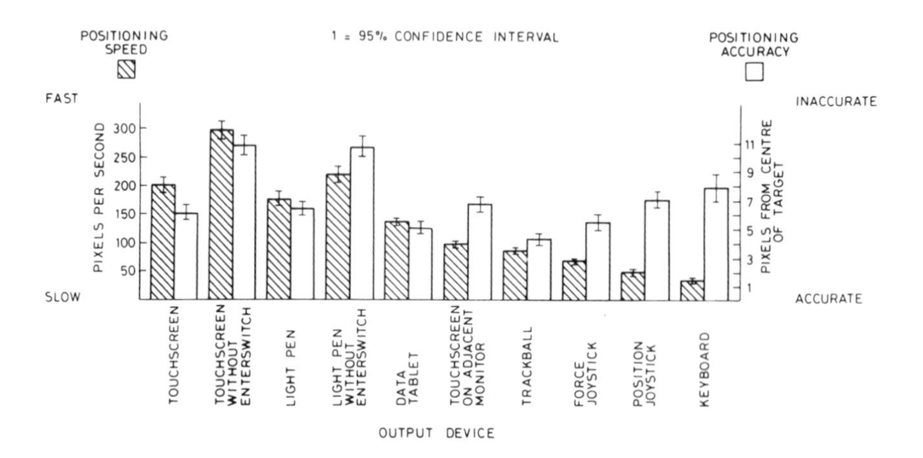

Fig. 6.13 — Comparison of devices (with acknowledgement to Albert 1982).

high as with monochrome screens even if it may be sufficient for many practical purposes. They are especially suitable for animated representation.

Newer forms of graphical storage devices such as *videodisks* and *compactdisks* allow, together with such media as touchscreens, the application of interactive videos and new forms of computer animation which combine computer and real images. They allow a large amount of pictorial data to be stored with very fast retrieval. For example, videodisks have a capacity from 40 000 up to 100 000 pictures. Videodisks combined with touchscreens are easy to use and show good results in interactive training cases (see Roberts 1985). Compactdisks, on the other hand, have slower retrieval times, but are well suited for the storage of single images.

Other output devices are *printers* and *plotters*. Printers are based on such technologies as dot matrix or laser, and work at high speed. Modern devices also allow graphics to be printed. There are also colour printers and photographic printers which produce slides or an output to microfilm. Plotters are suitable for high-quality graphic output and are also able to produce coloured output.

6.8 ADDITIONAL INFORMATION ON SCREEN DESIGN, COLOURS, ANIMATION, HELP FACILITIES AND ERROR MESSAGES

The following section describes interface features and tools for its design which have not been mentioned up to now.

There are some general rules which can be applied to *screen design* and which are important especially for interactive systems:

— Overloading the screen with too much information should be avoided;

where there are large amounts of output data, it is better to group them into several pages.
— Output lines should be justified on the left or right, and printing on the screen should accord with reading habits, i.e. from left to right and from the top down.
— The logical sequence of the computer tasks should be respected by the output, and the output should be structured semantically.
— Global and local screen complexity (the number of characters on the whole or on parts of the screen) should be kept low to enable fast comprehension.
— Groups of information should be separated from other groups by a blank area.

The use of *colour* can improve performance significantly and may draw the user's attention to specific points. It gives the additional possibility of drawing logical linkages and associations between data. It provides a more natural form of output, which is more easily recognized by the user, and it can respect the user's emotions. It is difficult to give general guidelines, as they depend very much on the specific application, but the following can be mentioned:

— Limit the use of colours and do not overload.
— Colours are a coding technique and they should draw attention to correlated data.
— Design an output for a monochrome screen first, thus keeping the logic more concise.
— Consistency in the use of colours has to be respected, for example use the same colour on all pages for the same semantic data.
— Attention should be paid to the fact that colour displays normally have a lower resolution than monochrome ones.

Windows (see Fig. 6.14) are a supplementary technique to structure the output, but they also serve for the input and the control of a program. They separate the screen into different areas, and often they are logically connected to different tasks. The use of windows is mostly combined with the use of a mouse as an input device.

A screen should have sufficient resolution to represent the single windows in a proper way. A mouse may support the moving, shrinking and closing of windows, and the overlapping of windows is also made possible. Furthermore, windows have the advantage of hiding information until it is explicitly called. They are a rather fast technique and also enable novice users to have satisfying interaction with a computer.

Animation graphics allow for the graphic representation of the actual state of program execution. The graphics are automatically updated with the development of the execution. This provides positive visual feedback for a user, it informs online, and it is useful not only for novices. As an example of a rather unusual case, Fig. 6.15 shows an animated representation of a sorting algorithm (Brown & Sedgewick 1987). One can distinguish the

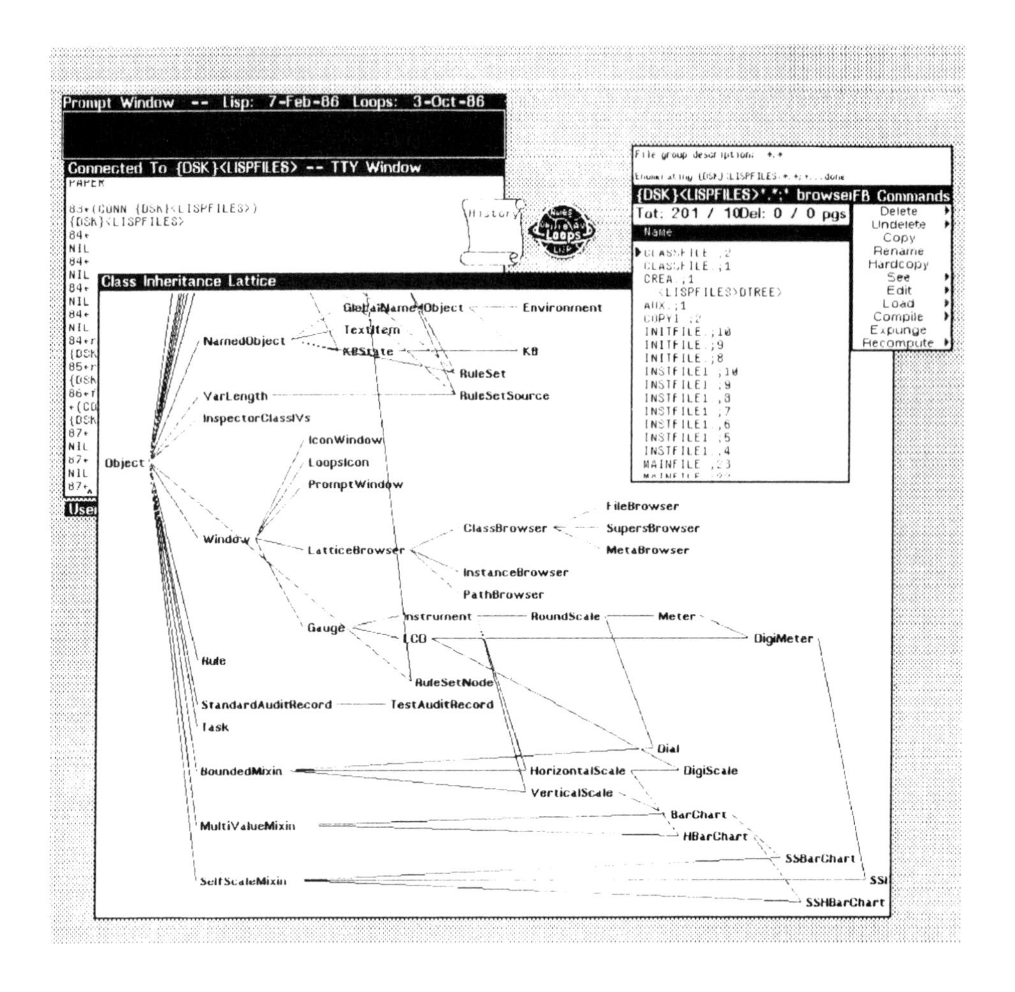

Fig. 6.14 — Windowing on a Rank Xerox 1186 workstation with an object browser.

sorting sequence from the left to the right and the different bars indicate the numbers to be sorted.

Another example is the representation of program objects on the screen. Fig. 6.16 presents a snapshot of a screen during a simulation, which moves cars, written in a SIMSCRIPT II.5 PC environment. All cars have as associated state variables their position and velocity, and every time these variables are updated, their graphical representations also change their positions on the screen automatically.

Manuals, help screens and tutorials are considered to be the main *help facilities*. Manuals in the form of written text that can be read before use are preferable for novice users. Clear entry points, a non-technical language and a task-specific grouping of information are necessary. On the other hand, online help facilities have an advantage in that they are available during the use of a computer and no searching in large volumes is necessary. Context-

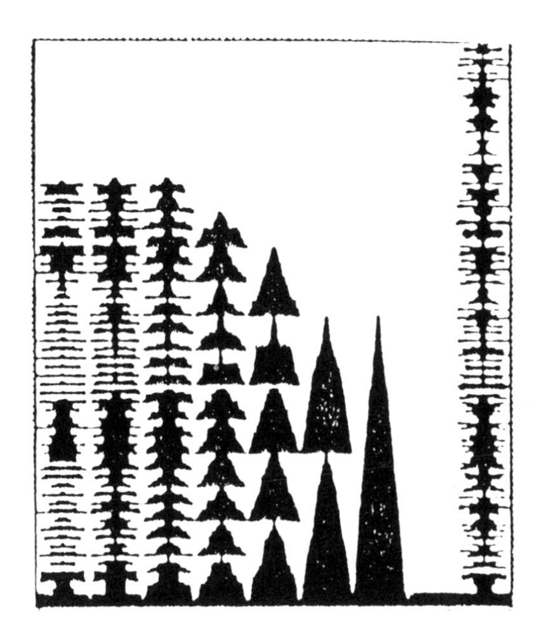

Fig. 6.15 — Animation of a sorting algorithm (from Brown & Sedgewick 1987).

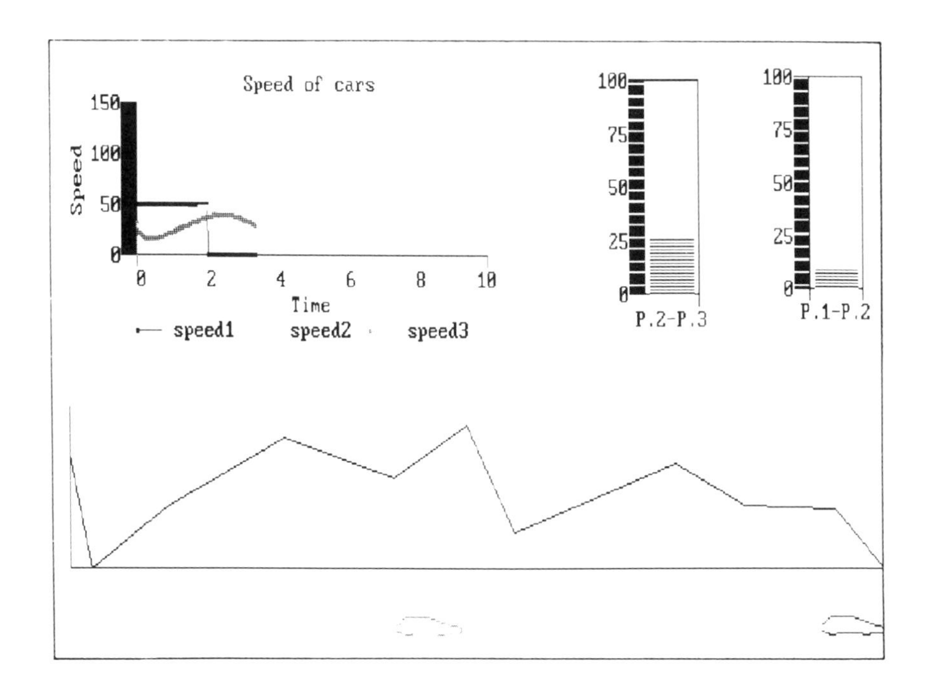

Fig. 6.16 — Animation with SIMSCRIPT II.5 (Simscript 1987).†
†Simscript is a registered trademark of CACI, Inc.

sensitive calling up of help screens may give the user the feeling of being permanently supported. The abilities of a computer to produce animation on a help screen may also support deeper understanding. But one should consider that written text is easier to read, and more information can be given on a paper page than on a screen. Also, a novice user may have problems in calling the help facility in a correct manner. The division of the screen into a part for help and a part for the actual work may also overload the screen. Full pages for help information may on the other hand put some stress on the short-term memory of the actual work, which is not seen during the study of the help page. Online tutorials serve mainly to demonstrate the syntax and semantics of a program before use, but they may also be called during the use of the program.

Finally, there are some guidelines for writing and evaluating *error messages*. Hopefully error messages will not be necessary, but, in case of need, they should be the first advice givers. Messages should be:

— Specific: a message such as SYNTAX ERROR is not enough; the user should be told what error has occurred where and advice for correcting the error should be given;
— Positive and constructive: examples such as FATAL ERROR or RUN ABORTED put stress on the user and hinder further usage;
— User-centred: a message has to give the user the feeling that he or she controls the system, i.e. it should be friendly;
— In a consistent and comprehensible format: (AH016) IN ROUTINE CUR_XEND ERROR is a negative example of a long numeric code, which is difficult even for experienced users to understand. Furthermore, error messages should be consistently located at the same place, for example in program compilations, always after the line where the error occurred, or alternatively all at the end of the program.

6.9 EXAMPLE 1: INTERFACE FOR A MODEL RETRIEVAL SYSTEM

The first example shows the use of a sophisticated interface in a model retrieval system (Guariso and Werthner 1988; see also Chapter 4). The system contains the description of water resources models available at a computer laboratory. The whole system is designed to enable a user, coming from industry or a university with an unknown level of skill and only a general description of his or her problem, to find a model which satisfies his or her needs. The system gives information on where one can find a model, how to use it, what limits one has to respect and what equipment is necessary to run it. The user types in several keywords to define the problem. With these words, which are translated to an internal database query, a screening process on the stored model descriptions starts. The system has two main parts, the model database and the user interface. The information in the modelbase is stored in a relational form. This information, which describes the models and which is subdivided into several relations, includes:

- keywords for the basic features of a model, such as its type, the scope of a model, necessary running environments, programming language, date of completion;
- the technique used and the structure of a model;
- computer-specific information, such as running on mainframes, minis or micros, operating systems, etc.;
- developers and suppliers, such as names of firms or institutes;
- extended documentation.

The first relation, which contains the main descriptive keywords, serves in the screening process to locate a model.

The design does not assume that the interface understands natural language input, although it enables the typing of single words in a free format and thus frees the average user from learning a strongly formatted DBMS query language. A deep understanding of natural language, i.e. understanding complete sentences with a broad range of words, is not supported by this approach. The interface implemented for the modelbase is limited to the handling of very simple English expressions, which describe some characteristics of the programs and are translated into the keywords used in the locating process. This approach is easy to fulfil in the proposed database structure, where only keywords play an essential role in the screening process. Another circumstance makes the design of the interface easier: it does not have to deal with the whole set of English words, but only with a relatively small subset of the language concerning software for environmental studies.

The user is allowed to type in his or her descriptive words and use some standard abbreviations, and the interface is also capable of detecting spelling mistakes. Another characteristic of the interface is its capability to learn from each user.

The interface of the retrieval system (see Fig. 6.17) consists of (a) a component which stores a dictionary of usable words, here the keywords and their synonyms, and (b) an analysing mechanism to interpret the user input and to learn from a new input.

DICTIONARY is a relation where the database keywords are stored one-to-one with their synonyms (see Table 6.4). Each keyword may have any number of synonyms or abbreviations, and both the keywords and their synonyms are stored in a standardized form (i.e. singular form, without special characters). The ANALYSER works on this separate relation DICTIONARY in a heuristic manner by looking for a synonym which matches the expression typed by the user.

The input may be composed of one or few words (normally nouns and adjectives), which are often used together to indicate a specific concept (e.g. water pollution, partial differential equations, etc.). First, it is standardized and partitioned, if it consists of more than one word. The whole input, in this standardized form, is then searched for among the values of the attribute

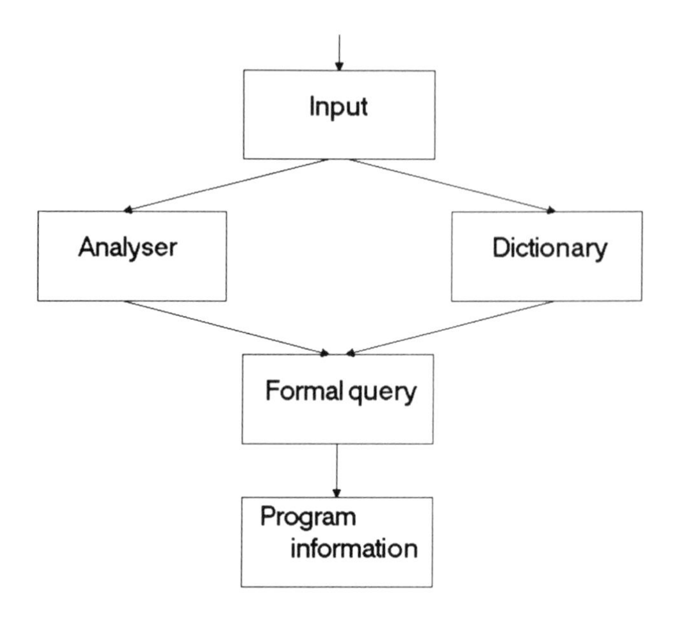

Fig. 6.17 — Structure of the interface.

Table 6.4 — Relation DICTIONARY with keywords and synonyms

KEYWORD (1st Attribute)	SYNONYM (2nd Attribute)
EQUATION	EQU
EQUATION	FORMULA
EQUATION	FORM

SYNONYM in the relation DICTIONARY. If the input matches an existing value, the search ends. If not, the system takes the first letter of the input (by this assuming that the probability of a correct first letter is very high) and a measure of the input length as basic criteria of similarity and tries to match all synonyms which fit these criteria. The comparison of the input and the synonym in the matching process also allows misspelling errors in the input to be detected. If conformity to a specified percentage is reached, the keyword is found, and the user, after a consistency check, is asked if he or she wants to insert his or her input as a new synonym into DICTIONARY.

If the search on the whole input fails and the input has been divided into single words, the process continues for each word, as described above. Each part, when found, is transformed into the related keyword. After locating all parts, the new transformed input (synonyms changed with the associated keywords) is searched for in DICTIONARY, again using the same procedure. (In the case of failure of the search, a list of all available keywords is displayed.)

Finally, the keywords are transformed into a formal query for locating the programs in the relation which contains the descriptive keywords for each of the models.

6.10 EXAMPLE 2: INTERFACE OF A QUALITATIVE MODEL BUILDER AND SIMULATOR

The interface of the qualitative model builder and simulator described in section 5.7 is presented below. In contrast to the previous one, which possesses rather traditional input and output characteristics, this interface is based on menus, graphical icons and windows; a mouse and the keyboard are used as input devices.

Entering the system, the user can choose whether to fetch an old model or to create a new one. The next menu allows the user to insert, delete or edit a variable. In the insert mode, the user is prompted for the name and type (input, state or output) of the variable. Afterwards, he or she places the graphical icon on the screen with the mouse, and then the different properties (see data structures in section 5.7) of the variable are asked for. If the user is not able to define some of the qualitative values, questions for ordering a variable with respect to the previously defined ones are prompted.

The system checks whether all necessary information with respect to the variable values is inserted. The user then has to define how the variables influence each other. This is done graphically by connecting two variables with the mouse. Furthermore the magnitude and sign of the connection is chosen from a menu. Thus the user creates a network of variables, as shown in Fig. 6.18.

This figure presents in two windows the qualitative system and the possibility for choosing the type of value for an input variable. The names and types of the variables and the sign of the connections are shown. A variable may be deleted by clicking the variable icon; for deleting a connection, the two related variables have to be marked.

After editing the model, which takes only some minutes, the user may pass to the simulation mode. There, he or she can define which variables are to be monitored during the run. The run time can also be set.

Fig. 6.19 presents the run time window. The clock indicates the number of time steps already computed. The monitored variables have analogue gauges which indicate the actual value using a qualitative scale (+extra big, +medium, 0, −medium and −extra big). Thus, the user is able to control the qualitative changes of the variables in the system. Since it makes no sense to present a numerical output in such a system, only the graphical representation is displayed. By clicking the mouse, the user may interrupt, set values, return to a backtrack point and then continue the simulation.

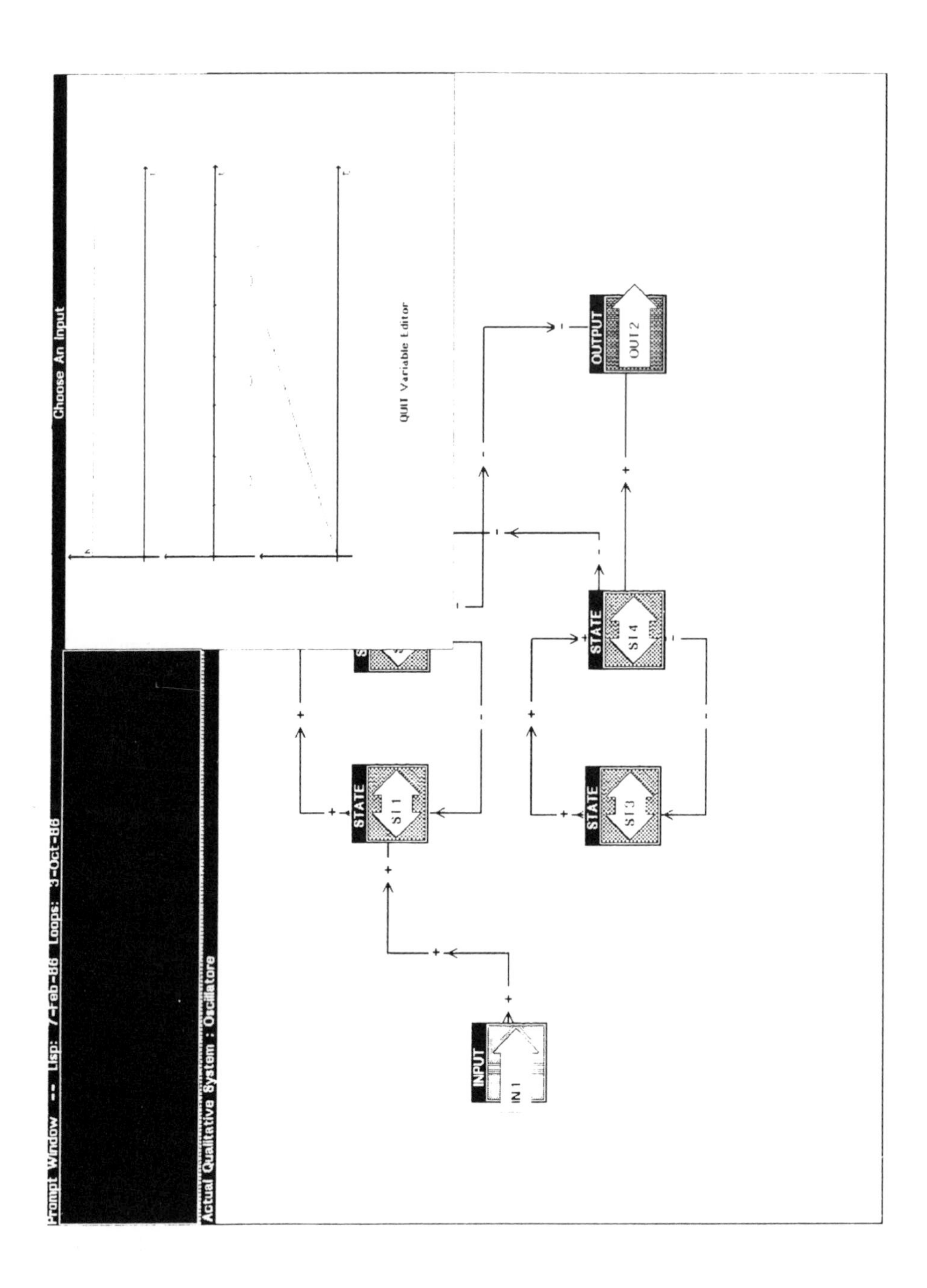

Fig. 6.18 — Editing of a qualitative model.

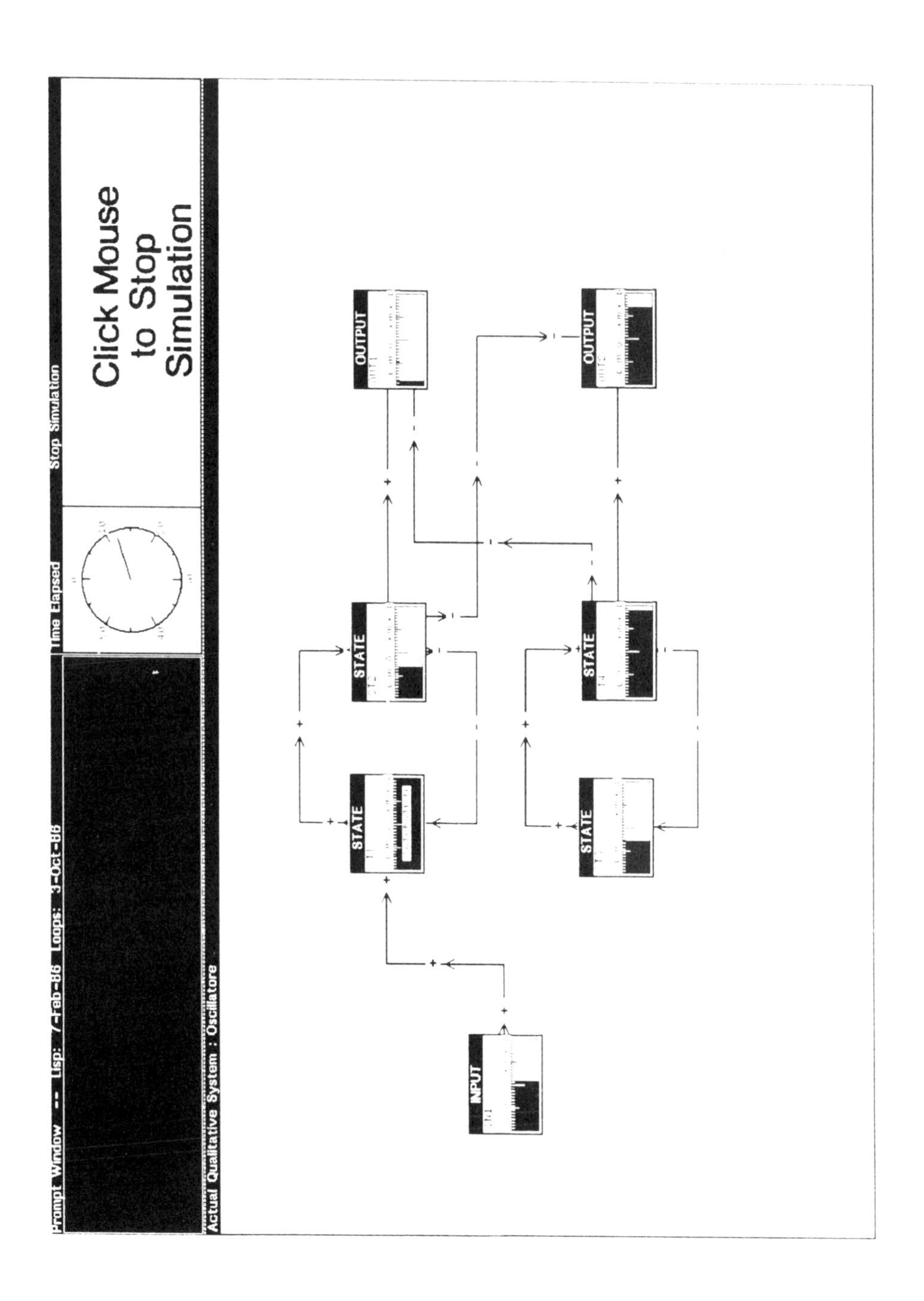

Fig. 6.19 — Simulation window with all variables monitored.

The extensive use of the system has shown that its design scope, i.e. enabling fast prototyping and simulation of the system, which can be described only in qualitative terms, has been reached. The interface design has proved to contain enough flexibility to satisfy the needs of the users and to be simple enough to limit substantially errors and stress.

REFERENCES

Albert, A.E. (1982) The effect of graphic input devices on performance in a cursor positioning task. In: *Proc. Human Factor Society — 26th Annual Meeting,* pp. 54–58.

Brighetti, S., Gandolfi, C., Gatto, M., Nardini, A., Rinaldi, S. & Werthner H. (1987) Experiencing the aid of a PC to teach ecology (in Italian). *Proc. Terzo Congresso Nazionale della Società Italiana di Ecologia,* Siena, Oct. 1987.

Brown, M.H. & Segdewick, R. (1984) A system for algorithm animation. *Computer Graphics* **18,** No. 3, 177–186.

Card, S.K., Moran, T.P. & Newell, A. (1980) The keystroke–level model for user performance time with interactive systems. *Comm. ACM* **23,** 396–410.

Chi, U.H. (1985) Formal specification of user interfaces: a comparison and evaluation of four axiomatic approaches. *IEEE Trans. Software Engineering* **11**, No. 8, 671–685.

Glasauer, H., Mattl, S. & Werthner, H. (1985) Computer in an exhibition about the Austrian Civil War 1934 (in German). *J. Sozialforschung* **25**, No. 4, 463–472.

Guariso, G. & Werthner, H. (1988) A software base for environmental studies. *Computer Journal,* **31**, No. 6, 550–553.

Guida, G. & Tasso, C. (1983) An expert intermediary system for interactive document retrieval. *Automatica* **19,** No. 6, 759–766.

Hammond, N. & Barnard, P. (1984) Dialogue design: characteristics of user knowledge. In: Monk, A. (ed.), *Fundamentals of Human–Computer Interaction.* Academic Press, London, pp. 127–164.

Hendrix, G.G., Sacerdoti, E.D., Sagalowicz, D. & Slocum, J. (1979) Developing a natural language interface to complex data. *ACM Trans. Database Systems* **3.**

Kiger, J.I. (1984) The depth/breadth trade–offs in the design of menu–driven user interfaces. *Int. J. of Man–Machine Studies* **20**, 201–213.

Kolers, P.A., Dachnicky, R.L. & Ferguson, D.C. (1981) Eye movement measurement of readability of CRT displays. *Human Factors* **23,** 517–527.

Miller, G.A. (1956) The magical number seven, plus or minus two: Some limits on our capacity for processing information. *Psychological Review* **63,** No. 2, 81–97.

Miller, R.B. (1968) Response time in man–computer conversational transactions. In: *Proc. Fall Joint Comp. Conf., AFIPS Press, Arlington,* pp. 267–277.

Moran, T.M. (1981) An applied psychology of the user. *Computing Surveys ACM* **13,** No. 1, 1–11.

Nake, F. (1984) Schnittstelle Mensch–Maschine. *Kursbuch* **75,** 109–118.

Neunteufel, R. (ed.) (1985) *Technology Assessment EDP. Consequences of the use of EDP on Labour, Economy and Society* (in German). Verlag des Österreichischen Gewerkschaftsbundes, Vienna.

Radl, G.W. (1980) Experimental investigations for optimal presentation–mode and colours of symbols on the CRT–screen. In: Grandjean, E. & Vigliani E. (eds), *Ergonomic Aspects of Visual Display Terminals.* Taylor & Francis, London.

Roberts, P.S. (1985) Intelligent computer based training. In: Johnson, P. & Cook, S. (eds) People and computers: designing the interface, *Proc. Conf. British Computer Society Human Computer Interaction Specialist Group 1988, University of East Anglia, 17–20 September 1985,* Cambridge University Press, Cambridge, UK, pp. 264–272.

Schank, R.C., & DeJong, G. (1979) Purposive understanding, In: Hayes, J.E., Michie, D. & Mikulich, L.I. (eds), *Machine Intelligence 9,* Ellis Horwood, Chichester, UK, pp. 459–478.

Shneiderman, B. (1987) *Designing the User Interface.* Addison–Wesley, Reading, Massachusetts.

SIMSCRIPT II.5 Programming Language (1987) CACI, Los Angeles, California.

Thomson, N. (1984) Human memory: different stores with different characteristics. In: Monk, A. (ed.), *Fundamentals of Human–Computer Interaction.* Academic Press, London, pp. 49–63.

Tinker, M.A. (1965) *Bases For Effective Reading.* University of Minnesota Press, Minnesota.

Tombaugh, J.W. (1985) The effects of VDU text–presentation rate on reading comprehension and reading speed. In: *Proc. CHI'85 — Human Factors in Computing Systems, ACM, Baltimore,* pp. 1–6.

Waltz, D.L. (1978) An English language question answering system for large databases. *Comm. ACM,* **21.**

Wasserman, A.I., Pircher, P.A., Shewmake, D.T., & Kersten, M.L. (1986) Developing interactive information systems with the user software engineering methodology. *IEEE Trans. Software Engineering* **12**, No. 2, 326–345.

7

A prototypical implementation of an environmental decision support system: EDSS-1

This final chapter presents in some detail the design and implementation of EDSS-1, a decision support system for environmental problems that includes most of the basic features outlined in the previous chapters. It may thus be viewed as a final proposal and practical summary of the concepts developed in the present work. Although several choices made in its development can be criticized and probably will be changed in different hardware/software environments, the experience gained in its development may constitute a useful reference for future studies and applications.

The key features of EDSS-1 are a frame-based representation of knowledge and an object-oriented approach in software development. Its main scope lies in the support of a user in building and using models in order to solve environmental problems in a fast and effective way. The user is freed from all the details about programming and managing the computer, so that he or she can concentrate on the conceptualization of the environmental problem and its solution. Most models commonly used to solve practical simulation, forecasting, planning, and management problems can be implemented into the system and the user is guided in finding those which may be most suitable for his or her needs and in utilizing them in a proper way. EDSS-1 thus constitutes an ideal software environment for rapid model prototyping and fast screening of alternative decisions and for evaluating their consequences on environmental systems. In its core it can also be seen as an advanced simulation model generator (Mathewson 1975, Spriet and Vansteenkiste 1982).

The system is directed to two different types of users:

- Experienced modellers interested in testing alternative model formulations, who are also able to define and edit a model written in terms of differential or difference equations.
- Novice users who can describe their problems in general terms (such as the components of the system they want to examine) but are not able to describe mathematically the dynamics of each component and thus must be guided in the choice of appropriate models.

The system software administers a Modelbase, a Knowledge base, a

Database, and a Dialogue module. Emphasis will be given in this chapter on the combination of the Model and the Knowledge base since this is probably the most interesting aspect of EDSS-1. This architecture allows the integration into a unique system of both simulation and optimization methods, and the combined use of artificial intelligence methodologies and mathematical models, thus exploiting the advantages of both approaches. The user interface will also be presented in some detail, since, as already quoted, it is quite an important aspect from the application perspective.

This prototypical EDSS has evolved from MoBase (Guariso *et al.* 1988). It has been implemented in an INTERLISP-D/LOOPS software environment (Stefik and Bobrow 1985, Bobrow *et al.* 1986) on a RANK XEROX 1186 workstation and takes advantage of the presence of two processors: a LISP processor managing the logical description, selection, and maintenance of models as well as the overall control of the system and the user interface; and a normal personal computer processor to run the environmental models that are written in classical programming languages such as FORTRAN and Pascal.

7.1 THE ARCHITECTURE OF EDSS-1

The software architecture of this prototype EDSS is that already presented in section 1.5 (see Fig. 7.1). A system manager software (System manager) interacts with four different modules: the Database, the Modelbase, the Knowledge base and the Dialogue module. Each base is responsible for specific capabilities and features of the system and a clear distinction between them has been drawn.

The Modelbase contains executable programs (the so-called 'numeric models'), which can be seen as a procedural description of simulation (or forecasting) environmental models. From a system theory point of view, these are structured input/output models (Zeigler 1984, Ören 1984), particularly dynamic systems, which can be described by ordinary differential or difference equations. The programs, which represent these models, can also be viewed as parametric programs, because a proper experiment with its necessary values has to be defined before running them. If models are linked together, input–output relations with units and time ranges also have to be defined. Typical programs in the Modelbase deal with such problems as air and water pollution, water and renewable resource management, planning of treatment plants, and other environmental control actions.

In this prototype, the Knowledge base serves as the central part of the system, which manages all the information necessary to utilize the models in the Modelbase and to integrate them into more complex models. This base thus includes the meta-information about the structure of all the other components of the system. More precisely the Knowledge base provides information for checking the consistency of the connections between models

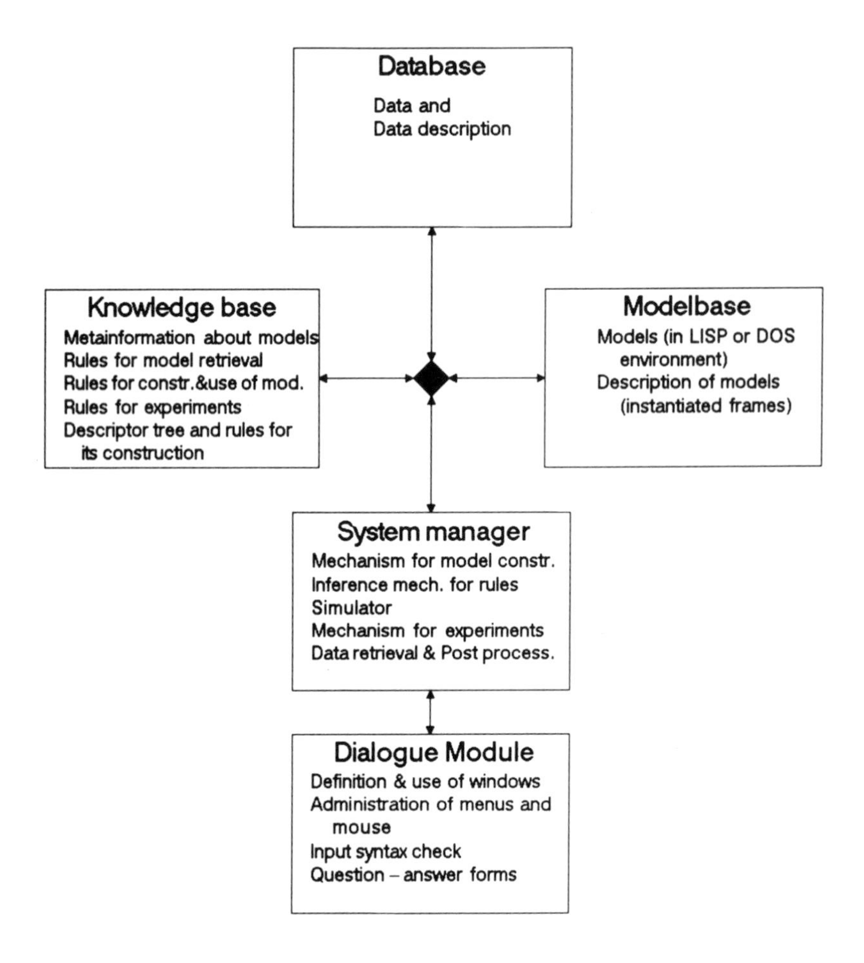

Fig. 7.1 — Main features of EDSS-1 modules.

and of the definitions of experiments, controls the sequence of simulations in case of multiple experiments, and provides advice on the type of model or model parameters to be used in a particular case.

The Database component of the system stores input and output values of experiments structured in terms of experimental frames, and provides data exchange for interfacing different components of a complex model. Default values for all model variables and parameters are also stored in this base.

The dialogue component manages input and output mainly via menus, fill-in forms and graphics. The user defines his or her models by filling in forms with a very simple and standard notation and connects graphically, using a mouse, the components of a model on the screen. After each experiment, the output may be post-processed in several ways, including basic statistical analyses.

The System manager contains all the procedures which implement the creation of new models, the retrieval of existing ones, and the execution of an experiment.

The main advantages of this architectural approach have been outlined in section 1.5:

— Gathering knowledge into a separate component makes it easier to maintain it and to add new knowledge.
— The distinction between the Model and the Knowledge base reflects in a proper way the necessary skills in the use of environmental models. Methods of knowledge representation as first order-logics (Nilsson 1980) and frames (Minsky 1981) may provide some static guidance for model use or keep structural knowledge on model properties (see also Chapter 5), whereas numerical models incorporate dynamic, time-dependent knowledge, which is typical of environmental systems.
— The presence of a unique interface module provides the user with a consistent view of the system activities in order to shorten the learning period and not put too much stress on his or her capabilities (see Chapter 6).

7.2 A FRAME-BASED REPRESENTATION OF MODELS AND EXPERIMENTS

The numerical programs contained in the Modelbase are represented in the system and accessed by the user through frames (see section 5.2), in which the various slots correspond to the system properties (name, variables, parameters, etc.). This representation supports the construction and use of models, allows several consistency checks to be easily performed, and permits the properties of a compound model to be inferred from the structure of its components.

The EDSS-1 Modelbase contains models of the type presented in section 4.3.1 that can be written in the general form

$$M = (T, U, Y, X, P, \lambda, \delta)$$

where

$T=$ $(t_0, \Delta t, t_N, v)$ denotes the time region — simulation starts at t_0 and proceeds in steps Δt to end at time t_N (measured in time units v);
$U=$ $\{u_1,\ldots,u_l\}$,
$Y=$ $\{y_1,\ldots,y_m\}$ and
$X=$ $\{x_1,\ldots,x_n\}$ denote the sets of input-, output- and state variables, respectively;
$P=$ $P_\lambda \cup P_\delta$ denotes the set of parameters of the functions λ and δ;
λ: $X \times U \times P_\lambda \times T \rightarrow X$ is the state transition function mapping state X_t to $X_{t+\Delta t}$ and
δ: $X \times P_\delta \times U \times T \rightarrow Y$ is the output function computing the values of Y: $Y_t = \delta(X_t, P_\delta, U_t, t)$.

This class of systems may be viewed as the root of a generalization hierarchy: by partial instantiation, i.e. by specifying the structures of U, Y, X and P and defining the functions λ and δ, it is partitioned into subclasses $\{B_1^*,..,B_r^*\}$ of so-called '(generic) basic models'. For each of them, a corresponding program that embodies the definitions of λ and δ must exist in the modelbase. Such a program is able to produce Y once T, U, initial values for X, and P are specified. Thus, the specification of these parameters instantiates a generic basic model B^* to a distinct basic model B (see also Fig. 7.4).

Memoryless (or static) models (Ören 1984) may be considered as special instances of M where no state variables are specified and the values of output variables are functions only of input variables, parameters and time:

$$Y_t = \delta(P, U_t, t)$$

The class of models considered is represented by uninstantiated (generic) frames. According to the system properties defined above, the generic frame (see Fig. 7.2) contains the slots INPUT VARIABLES, STATE VARIABLES, OUTPUT VARIABLES, PARAMETERS, and TIME UNIT. To distinguish the three different types of models (basic, static and compound), a slot TYPE (which is exclusively controlled by the system) is used. In addition, the slot CLASS is used to distinguish between continuous and discrete time models (the former are discretized before simulation with a standard Runge–Kutta algorithm); DESCRIPTION (textual information about the model both in colloquial terms and in the form of keyword-descriptors), SOFTWARE (for basic models: reference to the corresponding entry in the Model base identifying the program which embodies the functions λ and δ) and ICON (defining the graphical representation of a model) are provided. Finally, the slot EXPERIMENTS provides the required link with the Database where input and output data are stored.

The majority of the system properties defined in the slots of the model frame can also be represented by frames. For instance, an experiment is represented by a frame with slots for a name, a description, a class (single or multiple), an initial time, a final time, initial state and parameter values and method. Parameters are again described by a frame containing a name, a description, and a value. The parameter values assigned in this experimental frame overwrite those entered directly in the model frame.

The user has thus the possibility of accessing all the items characterizing a model and/or an experiment by following a tree structure of frames. As will be shown in the final section, this path is presented to the user as a cascade of windows which may overlap on the screen since only the innermost is active at any time.

7.3 THE MODELBASE

The EDSS-1 Modelbase contains executable programs representing environmental systems. These programs may be of two types:

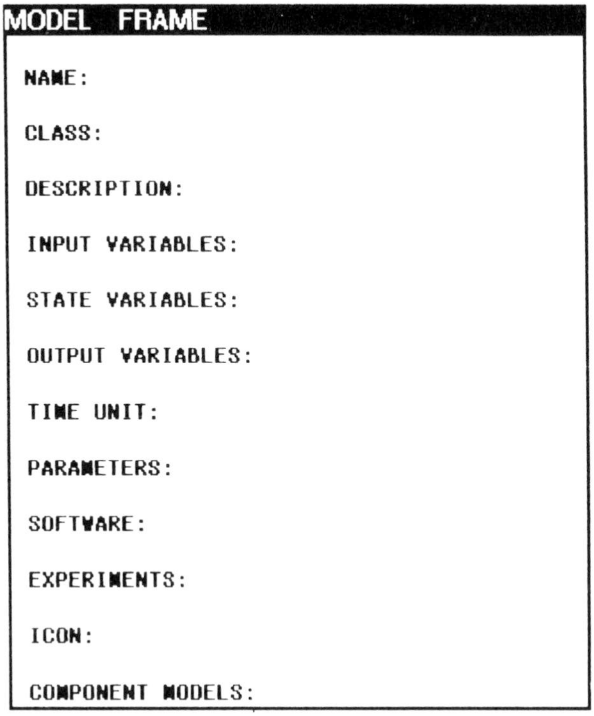

MODEL FRAME

NAME:

CLASS:

DESCRIPTION:

INPUT VARIABLES:

STATE VARIABLES:

OUTPUT VARIABLES:

TIME UNIT:

PARAMETERS:

SOFTWARE:

EXPERIMENTS:

ICON:

COMPONENT MODELS:

Fig. 7.2 — An empty model frame.

— written directly by a user with experience in modelling in terms of systems of equations, which can be completely accessed and modified and are finally translated by the system into executable LISP programs;
— programs written in any computer language which run under the DOS operating system on the personal computer processor of the workstation and can only be used without modifications.

If the models are used in their basic form, there is no difference between the two alternatives (the slot SOFTWARE in the model frame is left empty if the model runs on the LISP processor). However, EDSS-1 allows the definition of compound models built by linking input- and output-variables of component models (basic, static and also compound). In this case, if one or more component is a DOS program, the compound model must be free of loops. A circulation of information at each time step is in fact impossible when a model is executed outside LISP.

Each model in the base is described by an instance of the frame presented in Fig. 7.2 which thus constitutes the structured information for model storage and retrieval. For instance, the user has direct access to the frame of a certain model by calling its name and this action also retrieves the required model from the base.

For compound models, the slot COMPONENT MODELS contains the frames representing the component models and their linkage information. However, this slot is filled in by the inference mechanism of the System manager and is not accessible to the user.

The set of input variables of those component models that are not linked to output variables of some other component model still represents external input and therefore constitutes the set of input variables of the compound models. Its state variables and parameters are obtained by simply joining the state variable and parameter sets of the component models; whereas for output variables, the user may choose between two possibilities: either to collect all variables or to restrict the set to those not linked to other components. In the first case, output variables linked to input variables represent at the same time output variables of the compound model. As an example of a compound model and for the work with EDSS-1, Fig. 7.3 presents a model to control the development of an animal population described by a Leslie model.

7.4 THE KNOWLEDGE BASE

In the present prototype the Knowledge base has a central role. It has several purposes:

- it provides the meta-information about all the knowledge representation structures (mainly frames) in the system;
- it contains a rule-based system to support model and parameter choice;
- it contains the rules to guide the procedures of the System manager for performing multiple experiments; and
- it has a set of descriptors, organized in the form of a classification hierarchy, in order to localize a specific model.

7.4.1 Meta-information about models

As to the first aspect, the main features of this module are:

- Representation of the logical structure of models as described in section 7.3, i.e. the Knowledge base stores the meta-frames describing the model properties, which are instantiated each time a new model enters the Modelbase.
- Storage of information necessary for the construction of new models. In particular, this base provides the rules for linking submodels into a compound one: for instance, which types of variables can be linked together; how their unit of measurement must be converted when necessary; how to relate submodels which are defined on a different time step.
- Storage of information to perform consistency checks on models and user inputs, and of the inference procedure to derive the frame of a compound model from those of submodels.

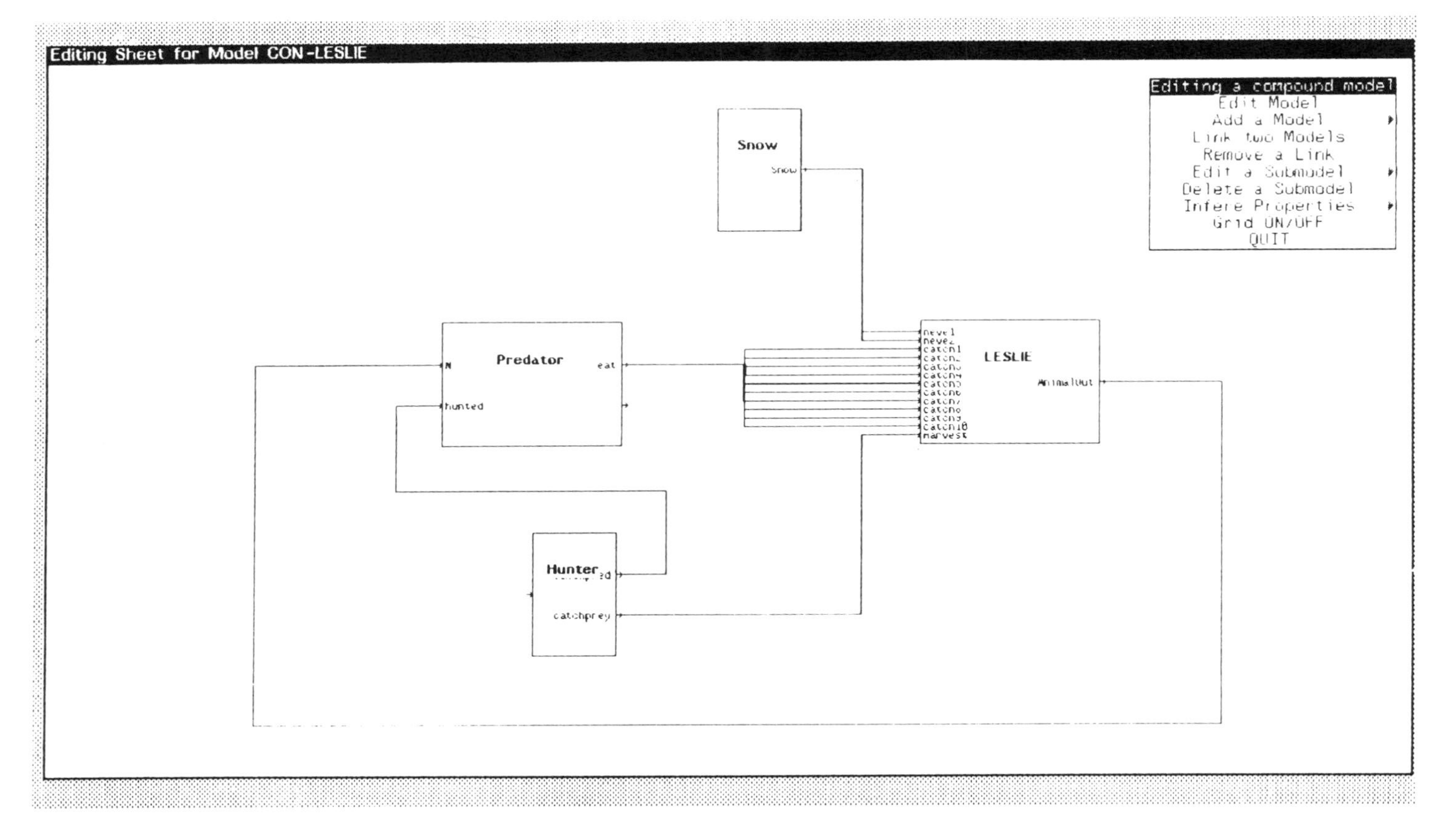

Fig. 7.3 — Graphic editing window for compound models.

The presence of basic, static, and compound models in the Modelbase can be viewed as the hierarchy shown in Fig. 7.4.

The generic class M of models (level 0) is subdivided into the three subtypes B (basic models), S (static models) and C (compound models) which are connected to M via an is-a relation (level I). B in turn is partitioned into the classes B_i^* of those members of B that correspond to existing programs; C into subclasses C_i^* consisting of compound models with the same structure; the same partition takes place for the static models S (level II). At the third level, the description of distinct models (instances of the types at level II) is stored in the Modelbase. Thus, the hierarchy of Fig. 7.4 combines both the Knowledge and the Modelbase. The upper part, the information on the general structure of models and how they have to be constructed, is maintained in the Knowledge base. The instances of these 'meta-frames', the distinct description of distinct models, is stored in the Modelbase. The entities at the lowest level can also be viewed as the experiments related to each model (for the relationship between experiments and stored data in EDSS-1 see the next section). This is mainly for implementational reasons and speeds up the model retrieval and usage process.

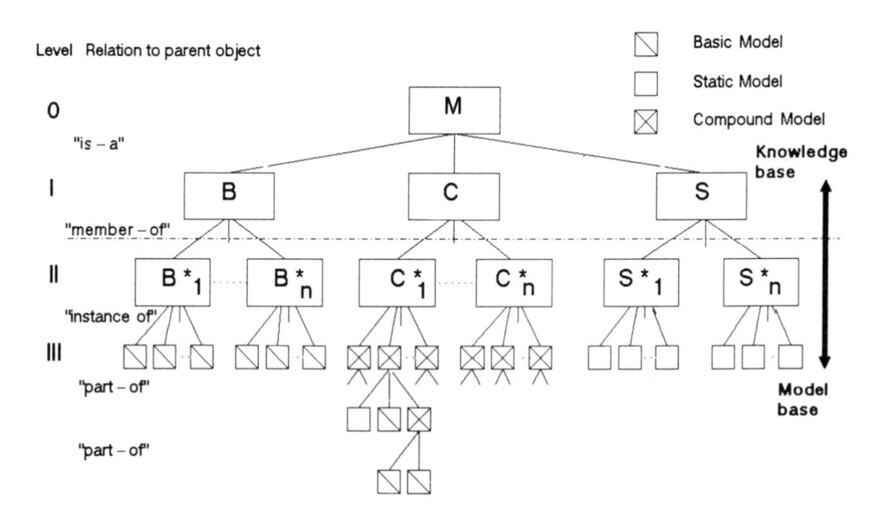

Fig. 7.4 — The model hierarchy.

The Knowledge base gives the user two supplementary ways to access models, besides directly entering their name: the first is to enter a certain number of descriptors of the features of the problem at hand or of the required model; the second is to enter a rule-based system which may advise the user about a satisfactory model choice.

7.4.2 A descriptor tree for finding a model

The main purpose of the descriptor tree is to guide an inexperienced user in finding a suitable model by allowing him to search the required model characteristics on a tree of model descriptors.

Any type of model (simulation, forecast, planning, and management; basic, static, or compound) can be catalogued with the help of such a tree. Each model is an object in the LOOPS environment with descriptors (keywords) associated with the DESCRIPTION slot in its frame.

The tree of descriptors serves for locating a model with respect to the user's need. The distinct descriptors are all instances of one LOOPS class (class DESCRIPTOR) and are linked by pointers. The descriptors are organized in a hierarchical manner allowing the user to find a model in a repetitive question–answer dialogue starting from a general point of view and arriving at a very discriminating one. These descriptors can be of two kinds: XOR nodes and OR nodes:

— XOR nodes allow the user to choose only one successor and represent semantically the choice of excluding alternatives;
— OR nodes let the user choose one or more successor nodes (descriptors) and correspond to different aspects of the model description.

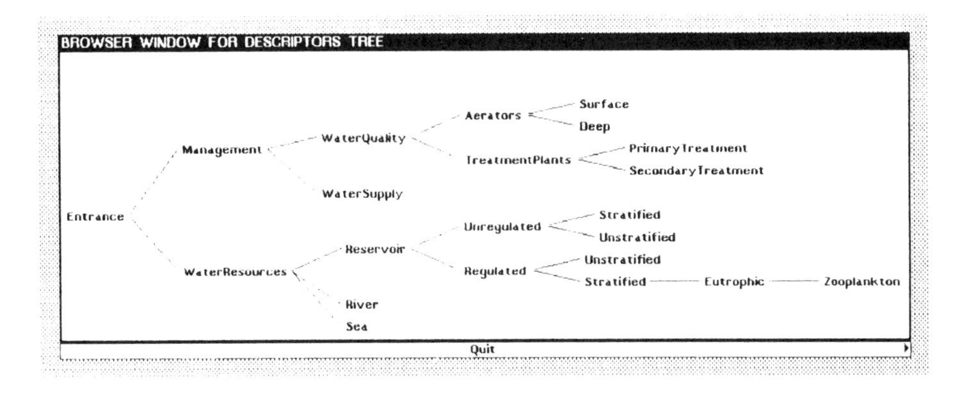

Fig. 7.5 — Sample OR/XOR tree for model description.

The tree of Fig. 7.5 presents a possible subtree of the overall descriptor structure. The user has the possibility of choosing between several main entrance points in the descriptor tree, i.e. aspects for describing his problem (application area — management, water resources; methods used; data types; etc.). The leaves of the tree contain one or more pointers to models in the base. A path through the tree is unique as it corresponds to one way of qualifying model attributes. But this path may correspond to different models, if they exhibit some common features. Since in many cases one path does not sufficiently isolate a model, the user may traverse the tree several times before a single model is selected.

The procedures or methods included in the LOOPS objects (particularly descriptors) provide also control on the user's actions. This control may take place at a local level (i.e. at each node of the tree structure) or at a global level.

Local control in the descriptor nodes (which is transparent for the user) allows the selection of just one or more successors depending upon the node type. Local control in the model classes (which form logically the leaves of the descriptor tree) means checking input data submitted by the user and calling the respective executable program.

Global control means, for example, guiding the number of paths in the descriptor tree until a model is discriminated, the detection of contradiction (for example, the set of models belonging to two descriptor paths is empty), justification of a model selection or interface control, and rules for inserting and deleting new descriptors into the tree structure.

7.4.3 A rule-based system to support model choice and use

In addition, support for model and parameter choice is provided by the rule-based system included in the Knowledge base. This is a simple expert system shell which uses LOOPS objects for the presentation of rules. Its inference engine works with a backward-chaining strategy to prove goals and is part of the System manager. These goals may be represented by a suggestion on the use of a certain model or certain parameter values. An example of the rules used for this purpose is given in Fig. 7.6.

Rules can be edited through a special editor and the inference mechanism which fires them can be visually followed on the screen. The consequence of a rule may also be an action such as asking the user for further details on his or her problem or directly implementing a suggested model.

Some meta-rules can also provide information on the structure of the knowledge in the base in order to scan only the part which is relevant to the problem at hand.

This way of using Expert Systems seems to be very promising in environmental problems and its application dates back to the first prototype developed by Gashing *et al.* in 1981. The system, called HYDRO, was in fact an intelligent advisor to choose the values of 16 parameters necessary to run a hydrological model (HSPF) developed by EPA. More recently this approach has been followed by Engman *et al.* (1986), Fayegh and Russell (1986), French and Baskar (1987), and Baffaut and Dalleur (1987). Possibly the best known of these systems is QUAL2E Advisor, an expert system developed at EPA by Barnwell *et al.* (1986) and presently under distribution. It helps in choosing suitable model formulation and input values for the water quality model QUAL2E, one of the most widely used packages in this domain.

The integration of a rule-based system with mathematical models seems to be one of the most suitable ways to transfer all the knowledge which has been accumulated on a certain model by the developers and by preceding

```
RULE FRAME

NAME: Streeter-Phelps Rule

IF

  problem is riverquality

 AND

  pollutant is biodegradable

THEN

  use Streeter-Phelps

NAME: reaeration-coefficient Rule

IF

  model is Streeter-Phelps

 AND

  river slow

THEN

  reaeration-coefficient = 0.3
```

Fig. 7.6 — Sample rules for model and parameter choice.

applications to an inexperienced user. The expert portion of the system may also help in analysing the causes of the departure of the simulation result from real data. In this respect, the availability of a rule editor is essential to allow an incremental construction of this part of the Knowledge base, in which additional information gained from applying the model can be easily added.

7.4.4 Procedures for multiple experiments

The last function of the EDSS-1 Knowledge base is to store the rules necessary to guide the procedures of the System manager for performing multiple experiments. Three procedures were implemented:

— the first simply scans a list of parameter/initial condition values and retains the results which optimize the function defined in the experiment method (see section 4.5);

— the second subdivides a parameter/initial condition interval in a given number of equally spaced values and tests all of them;
— the third is a classical Powell method to optimize a function without calculating the derivatives.

It is worth noticing that, since the multiple experiment procedure is external to the model simulator (a part of the System manager), the satisfaction of constraints on the model variables is checked only at the end of each elementary experiment and thus computer time cannot be spared when a constraint is violated at the beginning of the simulation. To take advantage of this fact, one should in fact modify the model formulation within the Modelbase, which is not allowed by the present architecture of EDSS-1 and may also be extremely complex from the user's point of view. Furthermore, input values cannot be modified during a multiple experiment since this is not required for parameter estimation or optimization of system performance.

7.5 THE DATABASE

The slots for input and output variables of the various models represent references to the Database, where a variable is defined as a frame containing information about name, textual description, measuring unit and data of a variable. Data (timeseries) may be specified in several ways. For example, the user may enter a series of constants representing values at fixed time steps, a sequence of time–value pairs, or an explicit function of time.

When submodels are defined on different time steps, timeseries are linearly interpolated to the smaller unit. Functions defining input data timeseries may be written by means of arithmetic expressions including ALGOL-like IF–THEN–ELSE constructs, several predefined functions, and all arithmetic LISP-functions, and may include the special symbol 't' that will be bound to the current time when evaluated.

The Database component of the system also serves for interfacing different component models and for providing models with necessary data for execution. Input/output relationships of sequentially executed models (the coupled components of a compound model) are stored in the Database and the respective frames are responsible for access to the Database. In the case of a compound model, the first component model delivers the output vectors, implemented conversion programs serve for the proper updating and the following component model is supplied with the necessary input vectors. However, all the information relative to model linking is directly managed by the system outside the user's control. Also default starting values for the state variables and the set of parameters are kept in the Database, if the user does not define his or her own values.

Both types of data, timeseries for the input and output vectors and scalar data such as parameters or initial values, can be accessed by the user through the model and experiment frames for maintenance and updating. Moreover, direct access to the Database is provided when analysing the output of the most recent experiment. This gives the user the possibility of post-processing these results and generating new variables as combinations of computed ones. This allows the user, for instance, to evaluate and to show directly the difference between two variables, to perform some statistical analysis, to rescale variables, etc. This is true also for multiple experiments for which only the output timeseries which maximize the function associated with the experimental method (see section 4.3) is stored in the base.

The overall organization of the Database and the method of accessing specific data are illustrated in Fig. 7.7.

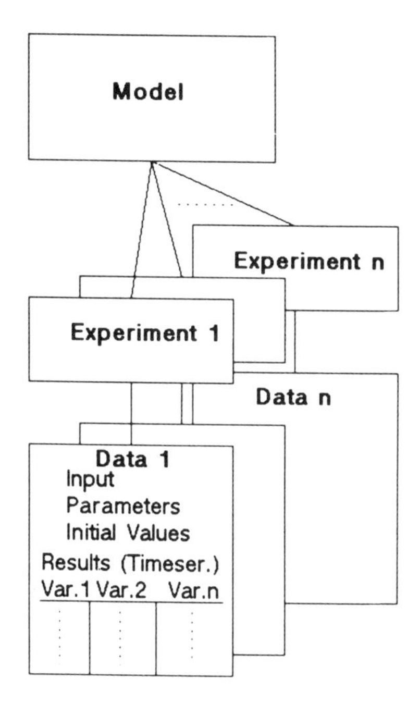

Fig. 7.7 — Structure of the EDSS-1 Database.

In EDSS-1, the Database structure is defined by LISP data objects and access mechanisms are implemented via LISP functions.

7.6 THE SYSTEM MANAGER

This component is constituted by a set of functions to access and manage the Model and Databases using the information provided by the Knowledge base. Furthermore, it implements the exchange of information with the user through the Dialogue Module.

More precisely, the system management software contains the simulator, the parser, the inference mechanism of the rule-based system and the control unit that performs the access to models, experiments and data, and the procedure which performs the construction of models.

The simulator works in different ways depending on the type of model at hand (basic, static or compound; DOS or LISP; continuous or discrete). Basic models are immediately simulated by the corresponding programs, compound models are simulated by a sequence of program calls, and static models are 'simulated' by a general purpose interpreter. DOS programs are executed in batch form on the PC processor (this means that all required data are prepared and transferred on a DOS file in the required format); LISP programs are directly executed after retrieval from the disk. LISP models written in a time-discrete form can be immediately simulated; time-continuous models are first discretized.

The simulation of a compound model is accomplished by running all the components in parallel (i.e. all the variables are recomputed at each time step) if all submodels have been defined in LISP; or by a coordinated sequence of programs in the opposite case.

During the aggregation process of compound models, several consistency checks are performed by the control unit:

— All data needed by the successor model must be generated by its preceding models.
— Variables that are linked together must be compatible with respect to their measurement units, which means that the unit of the output variable must be convertible to the unit of the input variable.
— Necessary data must be available in the Database.
— Functional expressions entered by the user must be correct.

The last capability is assured by the presence of a parser which analyses the expressions entered by the user and interprets them for formulating the basic model in the LISP environment and for static models. This allows the user to enter the functional definition of a model in the most common way, i.e. by writing down difference or differential equations exactly as they would be written on paper (see also the next section).

Finally, the control unit monitors all the activities of the system and provides connections between the bases and the interface module.

7.7 THE INTERFACE

A well-designed dialogue component is of major importance for acceptance of the system. EDSS-1 users, coming mainly from the field of environmental planning, may possess little knowledge of computer science. Via the dialogue part, the user has access to the different bases in an intuitive and consistent way.

In EDSS-1 input and output are performed mainly via menus, forms representing the frames and graphics. The user defines his or her models and experiments by filling in forms corresponding to the different frames. Whenever the number of choices is limited, the menu approach is used. Graphics are used not only for the traditional representation of the results but also to allow the construction of compound models. The user defines connections between submodels by using a cursor on the screen in such a way that the resulting model is represented by a directed graph. The cursor is driven by a mouse which allows choice and rejection.

Each form, menu and graph is represented by a window which can be closed, shrunk, moved, piled to other windows, or sent to hardcopy at any time during the work. Furthermore, the windows present only the information relevant to the user. For instance, the slot COMPONENT MODELS in the model frame of Fig. 7.2 is not shown by the associated window, since it is internally managed by the system.

The use of the keyboard is limited to the entering of numerical values and textual descriptions. Furthermore, immediate syntactic and semantic checks of the provided input are performed and the user is prompted with corresponding error messages in a dedicated, permanently displayed window.

In the construction process of a model the user gets an empty frame by clicking with the mouse at a specific menu and can fill in the respective slots. He or she can move from one slot to another with the mouse and type in names and numbers from the keyboard. For instance, when defining a state variable, a user gets the following sequence of windows: the model, the list of state variables, the single state variable with its properties and finally a window, where he or she defines the state transition function in a plain algebraic form. Only the second member of the system equations is entered since the first part is assumed to be either the derivative with respect to time for continuous models or the variable at the following time step for discrete ones. This window sequence is shown in Fig. 7.8.

Following this sequence of windows the user passes also the recursive storage structures of a specific model or experiment. The information can be thought of as organized in a hierarchic manner, at the top level the model, one step down the list of variables and so on. The immediately lower level of information is hidden from the user, but he or she can access and use it by simply clicking the mouse. By this approach, information hiding is provided together with system transparency and the user also gets a very concise idea of the system's working process. The same concept is used for the building of compound models. Submodels are linked by connecting their input and output variables to form a model on a higher level. The properties of the compound model are inferred from those of the components and the compound model becomes a member of the Modelbase, thus complex topological structures can be built. Fig. 7.9 is an example of this feature of EDSS-1.

The models are linked directly on the screen. The user selects models, places them on the screen and connects them by clicking with the mouse the predecessor and the successor model of the model sequence. Furthermore, he or she defines, with the help of the mouse, which input variable has to be connected to which output variable. The system automatically performs the update of the overall model and checks also whether the variables are compatible (ranges, dimensions) or not. Since input and output variables

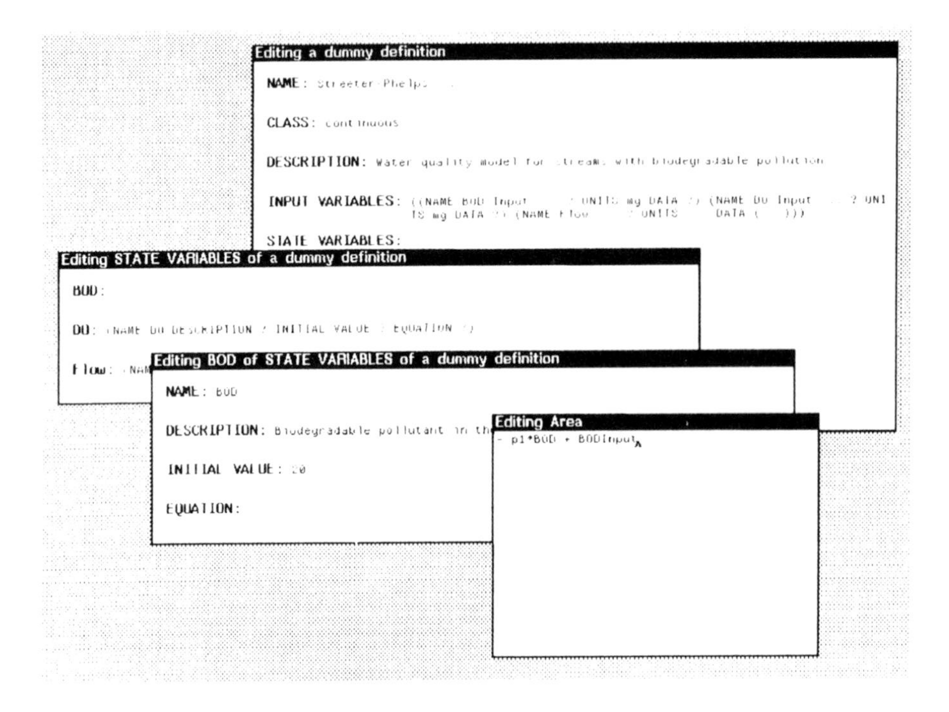

Fig. 7.8 — Sequence of windows to define (or modify) state variables and their transition functions.

may be expressed by functions containing IF–THEN–ELSE statements, it is also possible to realize time-dependent links that may be activated only when a variable satisfies a given condition.

After running an experiment, EDSS-1 offers the user some features for analysing input and output variables. There are facilities (see Fig. 7.10) for the following:

— browsing numerical values in a scrollable window;
— plotting variables versus time in different scales;
— computing aggregated information by means of statistical functions such as sum, mean, variance, autocorrelation, etc.; and
— defining new variables as functions of existing ones.

All information displayed on the screen may be dumped to the printer via an internal hardcopy facility.

7.8 AN EXAMPLE OF THE USE OF EDSS-1

A sample application of EDSS-1 is presented in this section to illustrate the use of the system. It refers to the problem of planning the operation of two waste water treatment plants which discharge their effluents into two

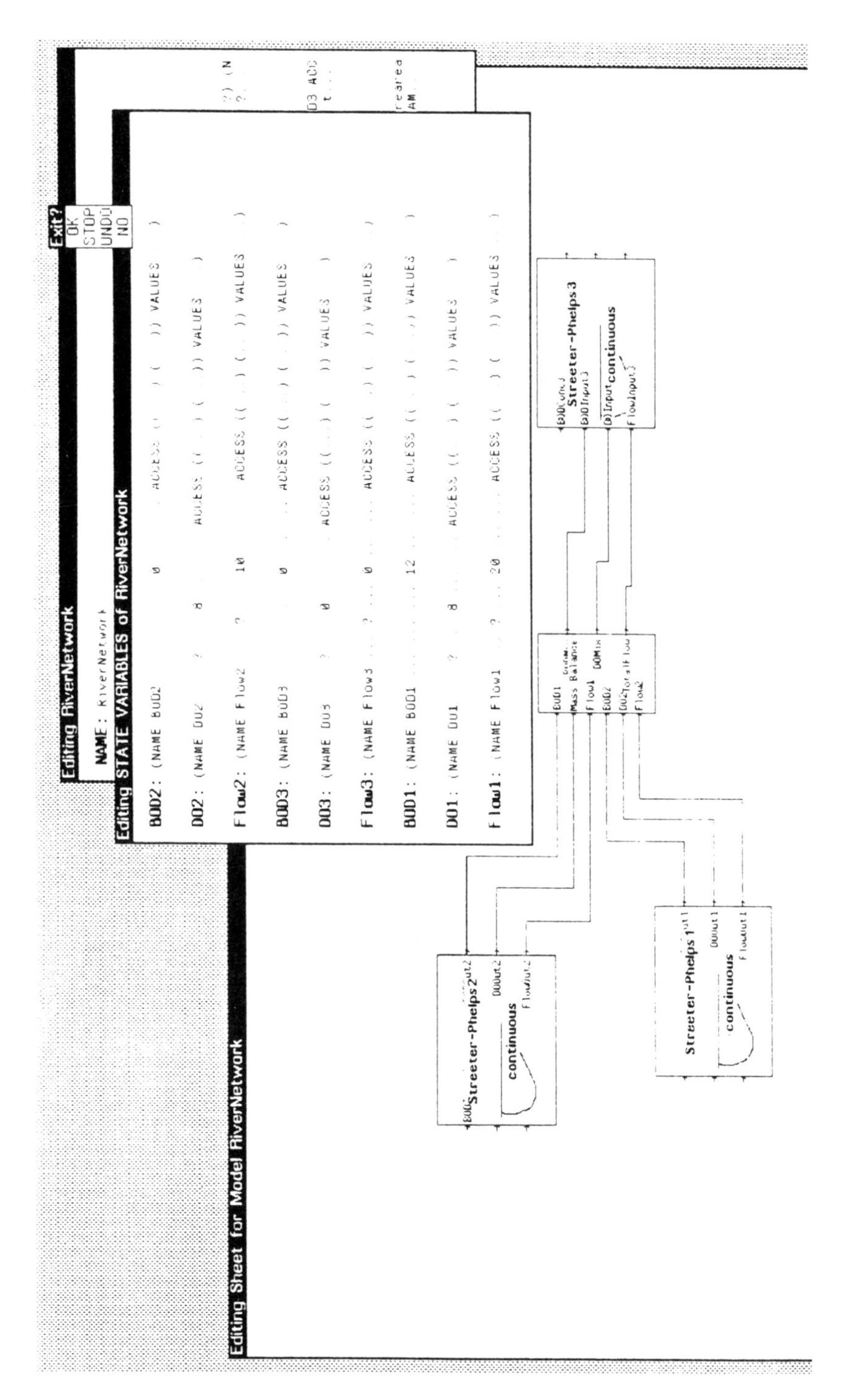

Fig. 7.9 — The River Network model is composed of three dynamic components of the third order and thus becomes a nine-state-variables system.

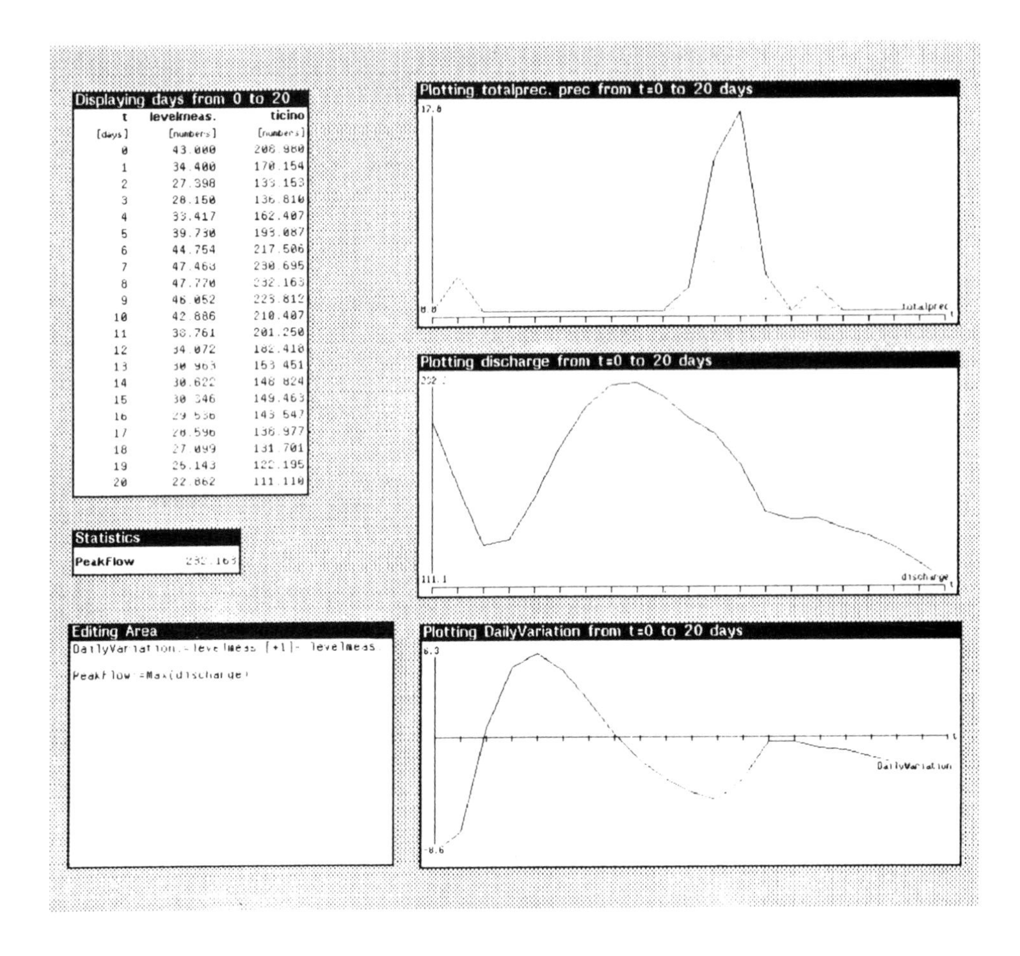

Fig. 7.10 — A snapshot of the output subsystem.

stretches of a simple river network. More precisely, the system must help in the choice of the level of treatment to be adopted in each plant in order to preserve a specified level of environmental quality while minimizing plant management costs. In this case, the minimum concentration of oxygen throughout the river network must be maintained above 4 mg/litre.

Three types of basic models were used for this study: the classical biochemical Streeter–Phelps model (see section 4.6.1); a static model representing the mass-balance equation at the junction of two river stretches; and another static model to represent the treatment plants.

The Streeter–Phelps water pollution model is a linear dynamic model with two state variables x_1 and x_2 representing respectively the biochemical oxygen demand (BOD) and the dissolved oxygen (DO) in the river. The state transition function λ of the model can be written as

$$\frac{dx_1}{dt} = -p_1x_1 + u_1(t)$$

$$\frac{dx_2}{dt} = -p_1x_1 + p_2(p_3 - x_2) + u_2(t)$$

with the addition of a third variable x_3 which represents the flow rate and which has been assumed constant along each river stretch:

$$\frac{dx_3}{dt} = 0$$

In these equations p_1 is the so-called degradation parameter, p_2 is the reoxygenation rate (the amount of oxygen taken from the atmosphere per unit of oxygen deficit), and p_3 represents the oxygen saturation concentration, which is a function of the water temperature and thus has been considered as a parameter. The two input variables u_1 and u_2 represent input of pollutants and of oxygen (this may be due to fresh water from the tributaries, or waterfalls or artificial aerators). In the three preceding equations, time is uniquely linked to distance by the law of motion of the river water, i.e. by the function $v(t)$, which expresses the velocity of the water versus time. Thanks to this, the model may be interpreted as expressing the state of a given particle of water flowing down the river, or the situation of the whole river in a stationary regime.

The mass-balance model is a static one, representing perfect mixing of two rivers at their junction. For x_1 and x_2 the mixing is a simple weighted mean of the inflows, i.e.

$$x_j^0 \frac{x_j^1 x_3^1 + x_j^2 x_3^2}{x_3^1 + x_3^2} \qquad j = 1,2$$

where x_j^0 represents the output values, while the superscripts 1 and 2 refer to the two joining rivers. As for the flow, the variable x_3, its output value is obviously the sum of the two corresponding inflows:

$$x_3^0 = x_3^1 + x_3^2$$

The compound model describing the junction of two rivers is represented by the system in Fig. 7.9, where the central block is the mass-balance equations and the other three blocks are Streeter–Phelps models for the three different river stretches. Using this compound model as a subsystem of more complex models, one can construct the water quality simulation model of any complex river network.

The last type of model represents the treatment plants, which are simplified to a proportionality relationship between pollutant loads and pollutant effluents. The cost function associated with each plant is given by a constant term plus an increasing (convex) function of the proportion of pollutant removed. This obviously means that the marginal cost of removing a unit of pollutant increases when required depuration tends to reach 100%. The structure of the overall system is represented in Fig. 7.11.

Also treatment plants have their own dynamics, and each plant could have been represented by a dynamic (compound) model. However, the problem at hand is to give a first idea on the operation of the two plants and thus, according to the approach developed in this book for EDSS, their internal structure does not seem to be relevant in this first phase.

If each plant is managed independently, each of them will solve a very simple optimization problem of the form

$$\min_{z} \text{ (treatment cost } (z))$$

subject to the constraint

$$\text{minimum oxygen concentration downstream } (z) < 4 \text{ mg/litre}$$

where the decision variable z represents, for each plant, the percentage of BOD load to be removed by treatment.

The solution to this problem, computed using EDSS-1, is represented in Fig. 7.12 and shows that obviously the treatments are set for both plants in such a way that the oxygen concentration exactly touches the minimum requirement at some point downstream of each plant.

The corresponding BOD concentrations are also shown in the upper part of Fig. 7.12.

Another possibility is that the two plants act in a cooperative way, i.e. there is a unique decision maker responsible for them who is interested in minimizing the total costs, while always satisfying the standard on minimum DO concentration. In this case, the problem can be formulated as

$$\min_{z_i} \Sigma_i \textit{ treatment cost}_i \ (z_i)$$

subject to the same constraint illustrated above.

Since the oxygen concentration of the water entering the third river stretch is not particularly high (around 7.5 mg/litre), a more intensive treatment at the first plant may increase this concentration and thus decrease the treatment required at the second plant. If the costs of the plant further upstream are lower, this policy may decrease the total costs of the system. Indeed this was the case in this study, and the solution of the cooperative problem is represented in Fig. 7.13.

The first plant treats more pollutant with an increase in its costs, but the downstream stretch receives cleaner water and thus the second plant may discharge more pollutant, thus reducing its costs to preserve the standard.

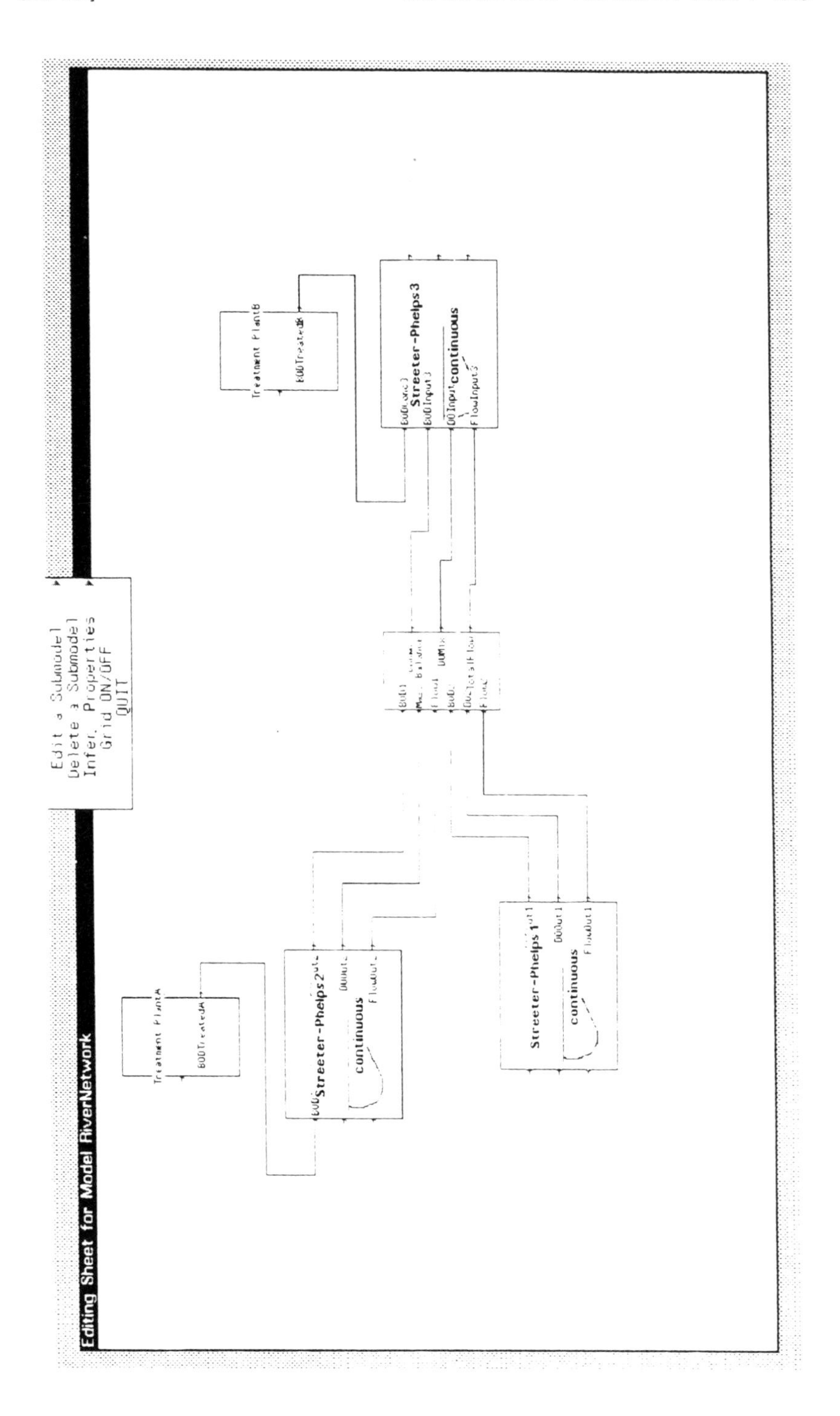

Fig. 7.11 — The model implemented to study the two treatment plants problem.

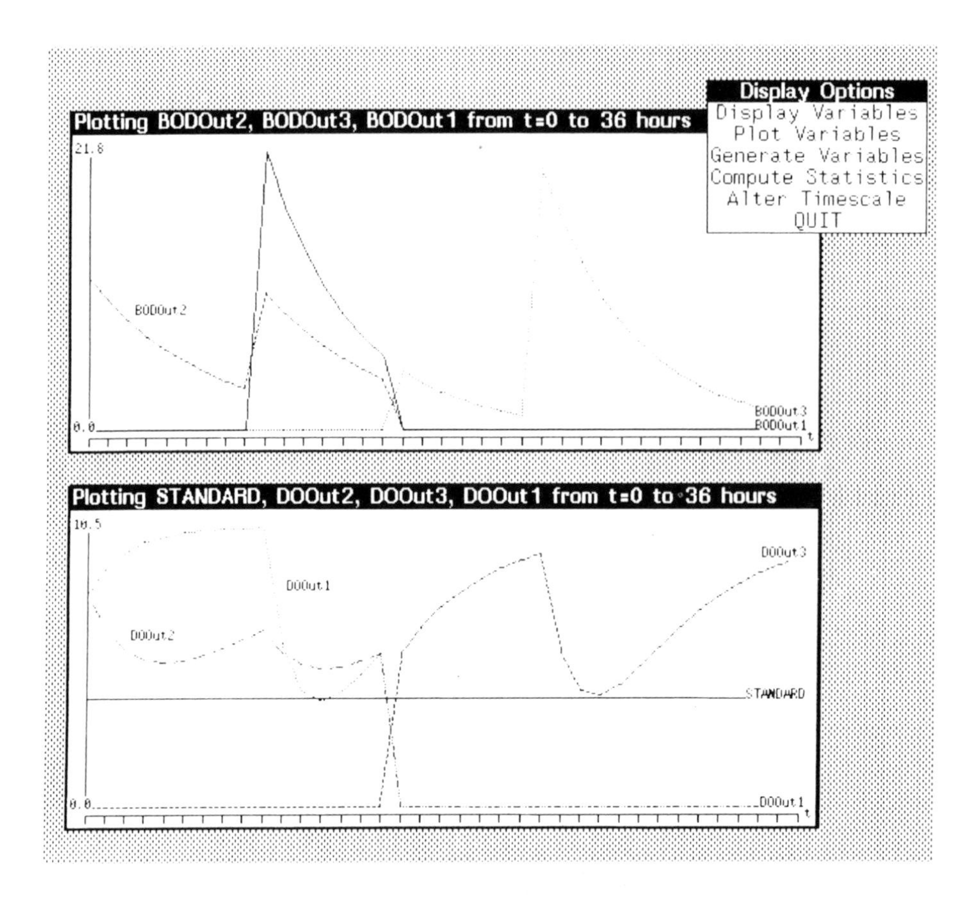

Fig. 7.12 — Optimization of the non-cooperative case: each plant treats the minimum amount of pollutant (upper figure) required to satisfy the oxygen standard (lower figure).

The optimal cooperative solution resulted, in this particular case, in a total cost reduction of around 10%.

REFERENCES

Baffaut, C. & Dalleur, C. W. (1987) Calibration of SWMM using an expert system. In: *Proc. IV Int. Conf. on Urban Storm Drainage, Lausanne, Switzerland.*

Barnwell, T. O., Brown, L. C. & Marek, W. (1986) *Development of a prototype expert advisor for the enhanced stream water quality model QUAL2E.* Internal Report, US EPA, Athens, Georgia.

Bobrow D. G., Kahn K. M. & Stefik M. J. (1986) CommonLoops merging Lisp and object oriented programming. *OOPSLA Proceedings* **86**, 17–29.

Engman, E. T., Rango, A. & Martinec, J. (1986) An expert system for

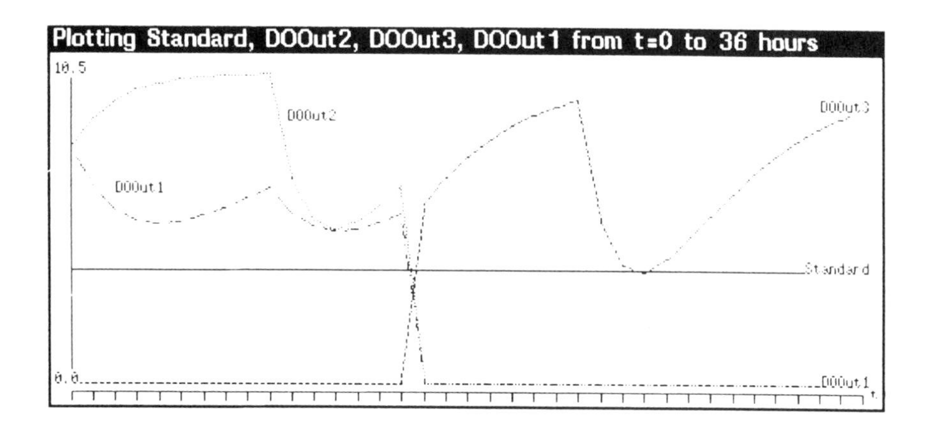

Fig. 7.13 — Optimal oxygen concentration in the three stretches in the cooperative case.

snowmelt runoff modelling and forecast. In: *Proc. Water Forum*, American Soc. Civil Engineers, New York.

Fayegh, D. & Russell, S. O. (1986) Flood advisor. In: Kostem, C. N. & Maher, M. L. (eds), *Expert Systems in Civil Engineering*. American Soc. Civil Engineers, New York.

French, N. M. & Baskar, N. R. (1987) An expert system for runoff hydrograph estimation. Presented at the AGU Spring Meeting, Baltimore, Maryland.

Gashing, J., Reboh, R. & Reiter, J. (1981) *Development of a knowledge-based expert system for water resources problems*. SRI Project 1619, SRI International, Palo Alto, California.

Guariso, G., Hitz, M. & Werthner, H. (1988) A knowledge based simulation environment for fast prototyping. In: Huntsinger, R. C., Karplus, W. J., Kerckhoffs, E. J. & Vansteenkiste, G. C., *Simulation Environments. Proc. of European Simulation Multiconference 1988, Nice, France, June 1988*, SCS, pp. 187–192.

Mathewson, S. C. (1975) Simulation program generators. *Simulation* **23**, No. 6, 181–189.

Minsky, M. (1981) A framework for representing knowledge. In: Haugeland, J. (ed.), *Mind Design*. MIT Press, Cambridge, Massachusetts, pp. 95–128.

Nilsson, N. J. (1980) *Principles of Artificial Intelligence*. Tioga Publishing, Palo Alto, California.

Ören, T. I. (1984) GEST — A modelling and simulation language based on system theoretic concepts. In: Ören, T. I., Zeigler, B. P. & Elzas, M. S. (eds), *Simulation and Model-based Methodologies: An Integrative View*. Springer-Verlag, Berlin, pp. 281–336.

Spriet, J. A. & Vansteenkiste, G. C. (1982) *Computer-aided Modelling and Simulation*. Academic Press, London.

Stefik, M. J. & Bobrow D. G. (1985) Object-oriented programming: themes and variations. *AI Mag.*, Winter 1985, 40–62.

Zeigler, B. P. (1984) System theoretic foundations of modelling and simulation. In: Ören, T. I., Zeigler, B. P. & Elzas, M. S. (eds), *Simulation and Model-based Methodologies: An Integrative View*. Springer-Verlag, Berlin, pp. 91–118.

Concluding remarks

In this book, the theories and basics of decision making were not discussed in much detail. The book did not enter into the topic of the axioms which lead managers in their decision making, nor were possible human interactions in those processes dealt with. First of all, there does not seem to be an overall and consistent theory which leads much beyond the presumption that decisions are based on rational insights, hard facts and completely accessible knowledge; and a broad reflection of the several theories which might constitute the basis for such a discussion would have exceeded the limits of the book. Secondly, the scope of the book is the development of a general framework of EDSS from a computer-science point of view. The book reviews the different theories and approaches to DSS and embeds them in the field of environmental management. The main emphasis has thus been on a description of the different tools, corresponding to the bases of the architecture that was developed here, which are necessary to improve the decision making process and the integration of these tools. This has been done by presenting basics which do not depend on distinct implementational details. The specifics lie in the conditions which affect the application of DSS in environmental management, for example the types of models which have to be included in a modelbase. Furthermore, the possibilities and advantages of incorporating several methods of AI into a DSS framework have been described in some detail. The proposed architecture, with its various bases, should satisfy the objectives and, by its modularity, should make it possible to respect both the different needs of managers and the distinguishable organizational contexts.

As a conclusion, some expected consequences of the influence of DSS and information technology in general (see also Huber 1988) will be presented. In a way, they can also be seen as the final product of developments presented in this book and describe the most probable conditions in which decision makers in environmental institutions will operate.

The number of applications of EDSS and, more generally, of information technology in environmental institutions will increase rapidly. This is not restricted only to traditional hardware devices. For example, parallel hardware architectures will become important in the field, as such machines fit well with some features of environmental problems (e.g. systems of differential equations). Furthermore, the possibility of connecting machines that are located at distant places will improve the work of institutions.

Software costs will remain the critical entry in EDSS budgets. The decrease in hardware costs will continue, and the basic costs for software development will become more economical, for example through the availability of prototyping environments. However, the needs of users will increase as they see the capabilities of such systems — *l'appétit vient en mangeant* (the appetite comes with eating) — and this will call for specific software designs which cannot be automated.

The number and variety of people involved in decision processes will increase. Systems will become more user-friendly, they will be closer to the user, and the time necessary to learn the features of a machine will decrease. Furthermore, it will be easier to access information. Thus, more persons will have the possibility of participating in a decision process inside an organization.

The number of meetings will decrease and the involvement and also influence of experts will increase. Since meetings can be better organized and thus more effective, their number will decrease. On the other hand, it will be easier for experts to be involved in decision processes. They will have better technical opportunities to present their proposals in a more understandable way.

Both types of organizations, centralized and decentralized, will benefit from this development. The collective memory of an organization will be improved. Centralized organizations can access stored information more easily and control the effects of solutions in a better way. On the other hand, different parts of decentralized organizations will know what is happening at other locations more quickly and this will improve cooperation between the parties concerned.

The effectiveness of decisions will increase. Decisions can be reached faster, and they will be better suited to a specific situation. Their effects can be better controlled. Owing to the possibility of reviewing past solutions to similar problems and accessing historical data, new situations can be embedded into the organizational context. Thus, by the use of EDSS the quality of the decision process will also increase.

The time to reach and to authorize decisions will decrease. Since there will be fewer meetings and both experts and managers can be integrated into the process, the necessary time for decisions will decrease. Managers, even if they have an aversion to such systems, will be able to learn very fast and to use directly such applications for their own benefit. Thus, the time needed to authorize decisions will decrease, since even high-level positions are involved in the preparation phase of a decision.

The importance of modern computer science techniques, such as the methods of AI, will increase. An architecture, such as the one proposed in this book puts a great deal of weight on computer science tools and their integration. Methods for the representation of knowledge and for inferring new knowledge will become more important. They serve both for the integration of the different bases and for the control of the system. Moreover, they may guide a user and give him or her substantial support in the use of the system.

Problem identification may be easier, and the basis of a decision may be more objective. Because it will be possible to access similar problem formulations, to review the information in the database, and to produce fast prototypical solutions, the formulation and definition of a problem will become easier. Since the basics of a solution can be formalized, stored and reviewed, it will become increasingly possible to criticize them and to propose alternatives. Since EDSS should have the capability of making the presumptions of a specific solution explicit, a decision will also be more objective. The use of a common language, supported by a common framework, enables the participation of different parties with different points of view. As a consequence transparency can be reached and the process of bargaining 'properly' for generating a consensus can be supported (Elzas 1984).

There can be no doubt that decisions will still be made by people, but systems as proposed here will improve this process. As an example, consider the system dealing with Lake Maggiore that was presented in Chapter 4. In this case, all participants (farmers, municipal officials, the energy agency) would be able to improve their benefits from the Lake Maggiore control programme, since the formulation of the problem and possible alternative solutions showed that a higher benefit for one party would not discriminate against another party. The original conflict was based on a mental model which did not represent objectively the real situation.

Institutions have to prepare themselves to conquer these new challenges. This means not only technical qualifications but also the ability to work in a cooperative and interdisciplinary manner. The complexity of the problem calls for both a technical and a social dimension in its solution. The qualifications of an organization must be reviewed and improved. This will also lead to a change in educational institutions; for example, sanitary engineers or ecologists will also have to learn about computers, and programmers and systems analysts will have to know the fundamentals of environmental problems. Educational institutions should thus enter this area at an early stage.

Technical systems will not and cannot do the work which remains to be done by humans. It is still their responsibility to be aware of the environmental situation and to cope with all the problems connected with it. People will not solve environmental problems simply because they use EDSS, but such systems will make it easier for them trying to come to the correct decisions.

REFERENCES

Huber, G. P. (1988) Effects of decision and communication support technologies on organizational decision processes and structures. In: Lee, R. M., McCosh, A. M. & Migliarese, P. (eds), *Organizational Decision Support Systems. Proc. IFIP WG 8.3 Working Conf., Lake Como, Italy, June 1988*, North-Holland, Amsterdam, pp. 317–333.

Elzas, M. S. (1984) Concepts for model-based policy construction. In: Ören,

T. I., Zeigler, B. P. & Elzas, M. S. (eds), *Simulation and Model-based Methodologies: An Integrative View.* NATO–ASI Series No. 10, Springer-Verlag, Berlin, pp. 133–181.

Index